DATE DUE

Handbook of Distributed Feedback Laser Diodes

For a complete listing of the *Artech House Optoelectronics Library*,
turn to the back of this book.

Handbook of Distributed Feedback Laser Diodes

Geert Morthier
Patrick Vankwikelberge

Artech House, Inc.
Boston • London

Library of Congress Cataloging-in-Publication Data
Morthier, Geert
  Handbook of distributed feedback laser diodes / Geert Morthier, Patrick Vankwikelberge.
     p.   cm.
  Includes bibliographical references and index.
  ISBN 0-89006-607-8   (alk. paper)
  1. Diodes, Semiconductor.   2. Semiconductor lasers.
  I. Vankwikelberge, Patrick.   II. Title.
  TK7871.86.M65   1997
  621.3815'22—dc21   97-4213
                                                       CIP

British Library Cataloguing in Publication Data
Morthier, Geert
  Handbook of distributed feedback laser diodes
  1. Solid-state lasers 2. Semiconductor lasers 3. Diodes, Semiconductor
  I. Title II. Vankwikelberge, Patrick
  621.3'661

  ISBN 0-89006-607-8

Cover design by Eileen Hoff

© 1997 ARTECH HOUSE, INC.
685 Canton Street
Norwood, MA 02062

All rights reserved. Printed and bound in the United States of America. No part of this book may be reproduced or utilized in any form or by any means, electronic or mechanical, including photocopying, recording, or by any information storage and retrieval system, without permission in writing from the publisher.
   All terms mentioned in this book that are known to be trademarks or service marks have been appropriately capitalized. Artech House cannot attest to the accuracy of this information. Use of a term in this book should not be regarded as affecting the validity of any trademark or service mark.

International Standard Book Number: 0-89006-607-8
Library of Congress Catalog Card Number: 97-4213

10 9 8 7 6 5 4 3 2 1

Contents

Foreword   xi

Preface   xiii

Chapter 1   Introduction to Fabry-Perot and Distributed Feedback Laser Diodes   1
   1.1   Historical Background   2
   1.2   Laser Diode Device Structure   4
   1.3   Operation of the Laser Diode   10
       1.3.1   The Basic Concept of Fabry-Perot Laser Diodes   10
       1.3.2   Optical Material Parameters   12
       1.3.3   Thermal Aspects of Laser Diodes   13
   1.4   Essential Laser Diode Characteristics   15
       1.4.1   Static Characteristics   15
       1.4.2   Dynamic Characteristics   17
   1.5   Use of Laser Diodes in Optical Communications Systems   19
   1.6   Dynamic Single-Mode Laser Diodes   21
       1.6.1   Short-Cavity Lasers   21
       1.6.2   Coupled-Cavity Lasers   22
       1.6.3   Injection-Locked Lasers   23
       1.6.4   Laser Diodes With Distributed Optical Feedback   23
   1.7   Organization of This Book   26
   References   26

Chapter 2   Rate Equation Theory of Laser Diodes   29
   2.1   Carrier Density Rate Equation   30
   2.2   Photon Density Rate Equation   31
   2.3   Phase Equations   32
   2.4   Introducing Noise in the Rate Equations   33
   2.5   Optical Gain and Absorption   36
       2.5.1   Bulk Materials   38

	2.5.2	Quantum Wells	39
	2.5.3	Strained-Layer Quantum Wells	41
	2.5.4	Gain Suppression	42
2.6	Some Well-Known Solutions of the Rate Equations		44
	2.6.1	The Static Side-Mode Suppression	46
	2.6.2	FM and AM Behavior	46
	2.6.3	Harmonic Distortion Characteristics	50
	2.6.4	Large-Signal Characteristics	52
	2.6.5	Power Spectrum, Linewidth, and Intensity Noise	55
2.7	The Influence of External Reflections		58
References			58

Chapter 3	Coupled-Mode Theory of DFB Laser Diodes		61
3.1	The Physical Processes Inside a Laser Diode		62
	3.1.1	The Electrical Process: Electrical Carrier Transport	62
	3.1.2	The Electro-Optic Process: The Carrier-Photon Interactions	62
	3.1.3	The Optical Process: Optical Wave Propagation	63
	3.1.4	The Thermal Process: Heat Transport	64
3.2	The Need for Simplification		64
3.3	Assumptions About the Modeled Laser Structure		64
3.4	Optical Wave Propagation		66
	3.4.1	Description of the Optical Field	66
	3.4.2	The Scalar Wave Equation	68
	3.4.3	The Langevin Force	69
	3.4.4	Reduction Toward the Coupled-Wave Equations	70
3.5	Discussion of the Coupled-Mode Wave Equations		75
	3.5.1	The Bragg and Maximum Gain Wavelengths	75
	3.5.2	Influence of Radiation Modes in Higher Order Gratings	76
	3.5.3	Longitudinal Rate Equations for the Optical Field	79
	3.5.4	The Instantaneous Optical Frequencies	81
	3.5.5	Comments on Spontaneous Emission	81
3.6	The Electrical Transport Problem		83
	3.6.1	The Carrier Rate Equation	83
	3.6.2	Current or Voltage Drive of the Laser	85
3.7	The Standing-Wave Effect in Gain-Coupled Lasers		86
3.8	Boundary Conditions		88
References			90

Chapter 4	Applying the Coupled-Mode Theory		93
4.1	Threshold Solutions for Simple DFB Lasers		94
	4.1.1	AR-Coated Index- and Gain-Coupled DFB Lasers	95
	4.1.2	The Stopband or Energy Gap	100

	4.1.3	DFB Lasers With Reflecting Facets	101
	4.1.4	The λ/4 Phase-Shifted DFB Laser	102
	4.1.5	Second-Order Index-Coupled DFB Lasers	104
4.2	Numerical Solutions of the Coupled-Mode Model	105
4.3	The Narrowband Approach for Solving the Coupled-Mode Model	106
	4.3.1	Threshold Analysis	108
	4.3.2	Continuous-Wave Analysis	109
	4.3.3	Small-Signal Dynamic Analysis	111
	4.3.4	Noise Analysis	112
	4.3.5	Large-Signal Dynamic Analysis	113
4.4	The Broadband Approach for Solving the Coupled-Mode Model	113
4.5	Derivation of the Rate Equations	114
4.6	Longitudinal Spatial Hole Burning	117
4.7	Coupling Coefficients for DFB Lasers	117
References	119

Chapter 5   A Closer Look at the Carrier Injection   123
5.1	Heterojunctions and Semi-Insulating Materials	124
	5.1.1	Heterojunctions	124
	5.1.2	Semi-Insulating InP	128
5.2	Carrier Leakage Over Heterobarriers	131
5.3	Carrier Injection in Gain-Guided and Weakly Index-Guided Lasers	134
5.4	Lateral Current Leakage in Index-Guided Structures	136
5.5	Parasitic Elements	140
5.6	Microwave Effects	143
5.7	Circuit Modeling of Leakage and Parasitics	144
References	145

Chapter 6   The Spectrum of DFB Laser Diodes   147
6.1	Amplified Spontaneous Emission	148
6.2	Side-Mode Rejection and Yield of DFB Lasers	155
6.3	Degradation of the SMSR by Spatial Hole Burning	159
6.4	DFB Lasers With Reduced Spatial Hole Burning	167
	6.4.1	Nonuniform Injection	168
	6.4.2	Special Index-Coupled Structures	169
	6.4.3	Gain-Coupled Lasers	173
6.5	The Wavelength Tunability of DFB Lasers	177
6.6	Measurement of the ASE Spectrum in DFB Lasers	179
6.7	Extraction of Device Parameters From the Spectrum	182
References	184

Chapter 7   The IM and FM Behavior of DFB Laser Diodes   187

7.1	Measuring the IM Response of Laser Diodes		187
7.2	Measuring the FM Response of Laser Diodes		189
	7.2.1	FM Measurements Based on Fabry-Perot Interferometers	189
	7.2.2	The Gated, Delayed Self-Homodyne Technique	190
	7.2.3	Characterization of Laser Chirp Using Fiber Dispersion	192
7.3	The IM Response		193
	7.3.1	The Subgigahertz IM Response	193
	7.3.2	The Spatial Hole Burning Cutoff Frequency	195
	7.3.3	The High-Frequency (>GHz) IM Response	196
7.4	The FM Response		198
	7.4.1	The FM Response of Fabry-Perot Lasers	198
	7.4.2	The FM Response of DFB Lasers	200
7.5	Lateral Spatial Hole Burning		210
7.6	Dynamics of Quantum-Well Lasers		212
7.7	Designing High-Speed DFB Lasers		215
References			216

Chapter 8	Harmonic and Intermodulation Distortion in DFB Laser Diodes	219
8.1	Measuring the Harmonic Distortion	220
8.2	Influence of the Relaxation Oscillations	221
8.3	Influence of Gain Suppression	224
8.4	Influence of Spatial Hole Burning	226
8.5	Influence of Leakage Currents	237
8.6	Dips in the Bias and Frequency Dependence of the Distortion	240
8.7	Relation With CSO and CTB	241
8.8	Designing Highly Linear DFB Lasers	245
References		248

Chapter 9	Noise Characteristics of DFB Laser Diodes		251
9.1	Measuring Noise Characteristics		252
9.2	FM Noise in DFB Lasers		254
	9.2.1	Frequency Dependence of the FM Noise Spectrum	254
	9.2.2	Methods for the Calculation of the FM Noise of Complex Laser Structures	255
9.3	Linewidth of DFB Lasers		259
9.4	Causes of Linewidth Rebroadening in DFB Lasers		261
	9.4.1	The Presence of Side Modes	261
	9.4.2	Gain Suppression	264
	9.4.3	Dispersion in the Feedback	266

9.5	Relative Intensity Noise of DFB Lasers		269
	9.5.1	Frequency Dependence of the RIN in Single-Mode Lasers	270
	9.5.2	Factors Determining the Low-Frequency RIN	271
9.6	Designing Highly Coherent DFB Lasers		273
References			275

Chapter 10 Fabrication and Packaging of DFB Laser Diodes 279
  10.1 Laser Diode Fabrication Techniques 280
    10.1.1 Liquid-Phase Epitaxy 281
    10.1.2 Molecular Beam Epitaxy 282
    10.1.3 Metal-Organic Vapor-Phase Epitaxy 283
  10.2 Grating Fabrication Techniques 284
  10.3 Packaging of DFB Laser Diodes 285
    10.3.1 Electrical Aspects 286
    10.3.2 Optical Aspects 287
  References 291

Chapter 11 Epilogue 293
  11.1 Trends in Optical Transmission and the Impact on DFB Lasers 293
  11.2 Future Directions in Design and Manufacturing of DFB Lasers 296
  References 298

Glossary 301

List of Symbols 303

About the Authors 305

Index 307

Foreword

The explosive development of optical communications in the past 25 years has to a large extent been made possible by the almost simultaneous invention of the low loss optical fiber and the semiconductor laser. This extraordinary conjunction of two quite different but revolutionary components has led to the development of optical communication links with a recently demonstrated record capacity of 2.6 Tbps on a single fiber over a distance of 120 km. It is hard to grasp such an amount of information transmission: Expressed in terms of 64-Kbps telephone calls this corresponds to the simultaneous transmission over a single fiber of 40 million voice circuits. This tremendous improvement in system performance was made possible by the advances obtained in optical fiber and especially in semiconductor laser technology. Whether one uses threshold current density or spectral characteristics as figure of merit, the improvement in semiconductor lasers has followed a curve even more impressive than that of the integration density of silicon chips. In this process the structure of the semiconductor laser has evolved from a simple GaAs pn junction diode with cleaved facets as mirrors, into an incredibly complex structure consisting of an extremely complex sequence of epitaxial layers, quantum well layers, heterojunctions, gratings, mirrors, etc. Technologically this was made possible by the development of epitaxial growth techniques such as MOVPE and MBE, which allow control down to the level of single atomic layers As this technology now enables one to build even more complex photonic integrated circuits in which lasers are combined with other optical functions, the need for a better and deeper understanding of the operation of the laser diode becomes more acute. The modeling and design of semiconductor lasers in general and the most important category, the DFB lasers in particular, is a daunting task indeed. One is faced with the combination of a semiconductor transport problem in very complex heterostructures together with an electromagnetic wave propagation problem in a complicated corrugated waveguide system, which is very large compared to the wavelength. If one adds to that the fact that the problem needs to be solved in 3 spatial and the temporal dimension, it is obvious that special skills and techniques are required to come to grips with the problem. In this book the authors have succeeded in synthesizing their enormous experience in dealing with this problem in such a way as to give the reader a new and deeper insight into the operation and design of DFB lasers. In view of the crucial role played by DFB lasers in the further development of optical communications and more specifically in the new area of wavelength multiplexing and switching, this book is a timely and important addition to the libraries of engineers active in this exciting field.

Paul Lagasse
March 1997

Preface

At present, distributed feedback (DFB) laser diodes are the preferred optical transmitters in most advanced optical communication systems. They possess a number of advantages over other laser types, such as their single-mode stability and their possible low-noise operation. However, early fabricated devices often did not exhibit stable or low-noise behavior or they became unstable at rather low power levels. It is only because a lot of research has been undertaken that the currently fabricated devices have superior characteristics over most other types of laser diodes. Indeed, since the first experimental demonstration of the operation of DFB laser diodes about 20 years ago, a lot of understanding of the complex physics that govern the stability, dynamic responses, and noise has been gained. It is largely owing to this increased understanding that single-mode DFB laser diodes with low noise, high modulation bandwidth, and high linearity can now be fabricated with a relatively high yield.

This book is intended to give a comprehensive description of the different effects that determine the behavior of a DFB laser diode. Emphasis is on developing a detailed understanding of DFB lasers and on the derivation of guidelines for their design. To this end, Chapters 1 to 4 deal with the device physics and how they can be modeled. Both a lumped rate equation model and a longtitudinal coupled-wave equation model are presented. The design is covered in Chapters 5 to 9, in which the different aspects of the laser performance (i.e., current injection efficiency, spectral stability, dynamic behavior, nonlinear distortion, and noise characteristics) are subsequently discussed. These chapters contain a large number of illustrations and have been written with the aim of providing clear explanations. Finally, Chapter 10 treats the fabrication and packaging of DFB laser diodes, while the epilogue gives an outlook on future DFB laser devices and their use in emerging high-capacity optical transmission systems. Although this book is focused on DFB lasers, much of the material is directly applicable to Fabry-Perot lasers.

This book should thus be of interest to researchers, engineers, and students in device fabricaton and design, optical-fiber communicatons, and any other field in which DFB laser diodes are used. Any person with a reasonable background in semiconductor and electromagnetic theory should be able to follow the text easily.

Most of the authors' knowledge and understanding about the topic is the result of several years of research performed at the Department of Information Technology of the University of Ghent. It is therefore our great pleasure to acknowledge here the director of this department, Prof. Paul Lagasse, and the group leader of the Optoelectronics Modeling and Characterization group, Prof. Roel Baets, for providing all the necessary means and opportunities for doing this research. We owe them much for the opportunities to collaborate and interact with many other researchers, both at the department and at several internationally recognized industrial and academic laboratories. It would be an impossible task to list all persons with whom we had stimulating discussions or interesting collaborations and who thus contributed to our own work. We do not wish to forget our early collaborators in the European RACE programs EPLOT and CMC, however: Jens Buus (Gayton Photonics), Bart Verbeek and Piet Kuindersma (Philips Optoelectronics Center), Chris Park and Richard Ash (HP-Ipswich), and Bernt Borchert from Siemens. François Brillouet and Jean-Luc Beylat (Alcatel Alsthom Recherche) gave us our first industrial assignment and more or less initiated the work on the linearity of DFB lasers. We also benefited from a fruitful collaboration on gain-coupled laser diodes with Professor Kunio Tada and Professor Yoshiaki Makamo (University of Tokyo).

We finally also wish to thank our family and friends for moral support and for taking care of our extra-professional and social needs. The authors especially thank Sofie for her great support and belief.

Chapter 1

Introduction to Fabry-Perot and Distributed Feedback Laser Diodes

More and more, light is used as a carrier for information or power. This trend is largely due to the development of lasers and in particular laser diodes. The applications of laser diodes are various. As a source of light power, they are used to pump solid-state lasers and fiber amplifiers, and they appear in laser printers and in optical disk storage devices. They can also be used in optical measurement equipment, such as interferometers and optical-fiber sensors. Other application areas include optical-signal processing, and optical communications.

There are several reasons for the strong interest in laser diodes. Laser diodes have a high external efficiency, often an order of magnitude larger than with other laser types. Moreover, this high efficiency can be reached with a very compact semiconductor device; typical dimensions are below 500 µm. Power can be supplied to the laser through electrical current injection. This current can be delivered with low-voltage semiconductor electronics. The optical output of the laser diode can be modulated directly through modulation of the injected driving current. *Intensity modulation* (IM) and *frequency modulation* (FM) are possible in this way. In addition, all the above aspects point toward the possibility to integrate laser diodes with other optical or electronic components in a single chip.

To exploit the possible applications mentioned above, laser diodes with appropriate device characteristics are required. Depending on the application, certain characteristics will be more important than others and minimum specifications will be required for these characteristics to turn a laser diode into an applicable device. Numerous laser diode device structures have already been designed in order to improve the static, dynamic, and noise properties of laser diodes. It is, however, difficult to include all good properties in a single device type. Therefore, laser devices are designed and optimized with a particular application area in mind.

One of the most important application areas for laser diodes is fiber-optic communications. Both in the transport network and in the access network, fiber-optic communication systems are becoming the most cost-effective solution to cover the increasing demand for bandwidth.

In the transport network, commercial *synchronous digital hierarchy* (SDH) multiplexing and cross-connecting equipment at 2.4 Gbps is available and 10 Gbps SDH equipment is under development. Moreover, trials are already going on with *wavelength-division multiplexing* (WDM) systems that carry about 10 optical channels modulated digitally at up to 10 Gbps over distances of several hundreds of kilometers [1,2]. Soliton transmission experiments are multiplying, showing impressive results with transmission distances of several thousands of kilometers at 10- to 20-Gbps data rates [3]. Further, increases in the data rate are under investigation with rates already exceeding 100 Gbps. Most of these systems make use of laser diodes in their transmitter modules. In heterodyne or homodyne coherent receivers, laser diodes can also be used as local oscillators.

To realize the continuously increasing performance demands for these communications systems, more and more stringent requirements are imposed on laser diodes. As such narrow-linewidth, low-threshold, high-speed, *dynamic single-mode* (DSM) lasers are required. Many of those systems requirements cannot be realized with simple *Fabry-Perot* (F-P) laser diodes. They require advanced dynamic single-mode *distributed feedback* (DFB) laser diodes.

Besides digital modulation schemes, analog microwave modulation of the optical carrier is also used. In the local loop, analog modulation schemes appear in cable TV (CATV) systems where hybrid-fiber coax and subcarrier multiplexing technologies are used [4,5]. Another area where analog optical communications are present is in cellular wireless communications, where remote antennas can be fed with the analog radio signal from a base station through microwave fiber-optic systems. For such optical microwave systems, the linearity of some of the DFB laser diode characteristics is imperative.

Achieving DFB laser diodes that meet the performance needs of modern optical communications systems requires a detailed understanding of those devices. It is the purpose of this book to give the reader a thorough understanding of DFB laser diodes. In this chapter, the concept of Fabry-Perot and DFB laser diodes is introduced. Before we turn our attention to DFB lasers, we will look at the traditional Fabry-Perot laser diode. Understanding it is essential to understanding the more complex DFB lasers, to which the remainder of this book is dedicated.

1.1 HISTORICAL BACKGROUND

In the early 1960s, it was suggested that semiconductors could be used as laser materials. This early work resulted in the demonstration of laser action across gallium arsenide (GaAs) p-n junctions by current injection at temperatures of 77K. Due to the single GaAs p-n junction, those devices were called homostructure laser diodes. They already used cleaved facets as laser mirrors, but their threshold current density at room temperature was very high (>50,000 A/cm^2). This was caused by a bad transverse (i.e., perpendicular to the junction) optical field confinement and the presence of lossy GaAs regions surrounding the gain region. In homostructure lasers, the gain region occurs as a thin light-emitting region on the p-side of the p-n junction. Its thickness is determined by the diffusion length of the electrons injected from the n-side in the p-side. It is in the range of 1 to 3 µm. Only a small refractive index perturbation (0.1% to 1%) is associated with the gain, and therefore only a weak thin-film index guide is induced. As a result, the light spreads too much in the lossy GaAs surrounding the gain region, and this leads to very high threshold current densities.

Toward the end of the 1960s, the problems faced by homostructure lasers were countered by the introduction of *double heterostructure* (DH) laser diodes. In the DH device, a semiconductor with a narrow energy bandgap is sandwiched between two layers of a wider energy bandgap semiconductor. The two heterojunctions help to confine the carriers in the central layer (usually called the active layer), where an efficient population inversion creates the optical gain. This central layer is typically 0.1 to 0.5 µm thick. At the same time, the double heterostructure forms a thin-film optical waveguide. Since the DH index step is large, typically on the order of 6% to 9%, this waveguide strongly supports the transverse confinement of the generated photons to the central layer, where the stimulated emission occurs. Moreover, the cladding layers only show a small absorption at the emitted wavelength(s). Therefore, DH lasers have much lower threshold current densities (<1,500 A/cm^2) than homostructure lasers.

During the 1970s these DH lasers with cleaved facet mirrors were further refined [6]. The threshold current at room temperature decreased and the far-field properties improved. Whereas earlier laser diodes used a simple stripe contact to confine the injection current laterally (see Figure 1.1), new ridge waveguide and buried heterostructure lasers were now being designed. Besides offering better lateral current confinement, these lasers realize better lateral confinement of the optical field to the area of stimulated emission. Such lasers also have lower threshold currents and show a better lateral mode control.

In the early 1970s work on distributed feedback lasers began. This work was part of the research toward DSM semiconductor lasers. The aim was to obtain lasers with tight mode control for the transverse (electric (TE) or magnetic (TM)), lateral, and longitudinal modes during high-speed direct modulation. Transverse mode control is based on differences in reflectivity and loss for TE and TM modes. Lateral mode control is achieved through appropriate waveguide design. For longitudinal mode control, different approaches exist, but they are all based on increasing the wavelength selectivity of the laser resonator. Considered approaches include cleaved-coupled-cavity (C^3) lasers, external-mirror or external-cavity lasers, short-cavity

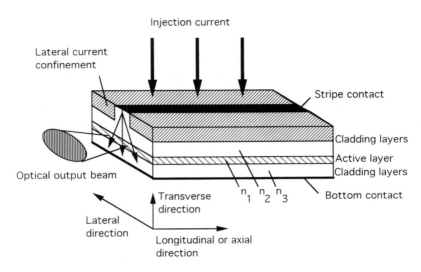

Figure 1.1 A simple stripe contact Fabry-Perot DH laser diode.

lasers, injection locking through light injection, DFB lasers, and distributed Bragg reflector (DBR) lasers. Among those alternatives, DFB lasers have prevailed as the most cost-effective approach. Secondary alternatives that are used today include external-cavity lasers and DBR lasers, while the remaining options have been dropped or only appear in very specific cases.

Although the theoretical work on DFB and DBR lasers started in the early 1970s [7,8], it was only toward the mid 1970s that DFB laser diode operation was demonstrated [9].

Since then, DFB lasers have evolved into highly advanced devices. The fabrication techniques, the DFB laser structures, and the understanding of the device physics have all greatly improved. New developments have, for instance, resulted in multi-section DFB lasers [10], multi-phase-shift DFB lasers [11], multi-quantum-well (MQW) DFB lasers [12], and gain-coupled DFB lasers [13].

In the meantime, DFB lasers have found their way to the market, where they have become a key component in advanced high-speed digital and analog fiber-optic communications systems.

1.2 LASER DIODE DEVICE STRUCTURE

As already mentioned before, the general structure and the basic operational principles of laser diodes will first be explained using the simple Fabry-Perot laser.

Figure 1.2 illustrates the principle of the Fabry-Perot laser. The laser cavity consists of two reflectors. Between them an active medium is present that amplifies a light beam that propagates forward and backward through the cavity. This cavity must also limit the diffraction of the propagating beam, possibly by means of an optical waveguide. The amplification of the active medium exists over the full length of the laser (typically a few hundred micrometers to 1 mm or more), and it is realized through pumping the active medium along the laser. In laser diodes the injection current takes care of this.

The most popular transverse laser diode structure is the DH laser diode, which can be represented in a simplified way by a three-layer semiconductor structure. Figure 1.1 shows a Fabry-Perot DH laser diode. The laser cavity is formed by the end facets of the laser chip. These are cleaved or sometimes etched and possibly covered with a coating. In between, the active optical waveguide is contained.

The term double heterostructure points to the presence of heterojunctions at both interfaces that separate the central active layer from the surrounding cladding layers. This central layer is the active layer in which the injected current creates the population inversion needed for the stimulated emission. As explained in the previous section, the cladding layers have a double purpose. First, they need to confine the light to the active layer. In this way, strong stimulated emission becomes possible in an efficient manner because the optical field has sufficient overlap with the recombining carriers in the active layer. To this end, the cladding layers and the active layer have to constitute a thin-film waveguide, which takes care of the transverse index-guiding. This puts constraints on the values of the refractive indexes of the layers $(n_1 > n_2, n_1 > n_3)$.

Second, a population inversion has to be created in the active layer. Therefore, a direct semiconductor is needed in the active layer. Moreover, the population

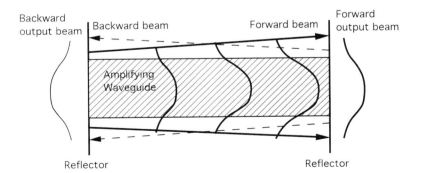

Figure 1.2 Basic concept of the Fabry-Perot laser: an optical amplifying medium in a resonator consisting of two mirrors.

inversion implies a sufficiently large concentration of electrons and holes in the conduction and valence bands, respectively, of the active layer. To realize those high carrier concentrations with acceptable injection currents (particularly with respect to the thermal load of the device), an efficient way of injecting carriers in the active layer is essential. However, it is important that the injected carriers do not leak out of the active layer. Therefore, the bandgap of the active layer needs to be smaller than the bandgap of the cladding layers. If the different layers also have the proper levels and types of doping, an efficient carrier injection is possible.

Figure 1.3 shows the energy band diagram of a laser diode (GaAs/AlGaAs or InGaAsP/InP) with zero bias and with a forward bias. The types and sizes of doping are indicated on the figure. Under forward bias, the two heterojunctions on either side of the active layer create potential barriers that prevent the carriers from leaking out of the active layer. The quasi-Fermi levels in this nonequilibrium situation show that the concentration of holes and electrons is high in the active layer, which is a

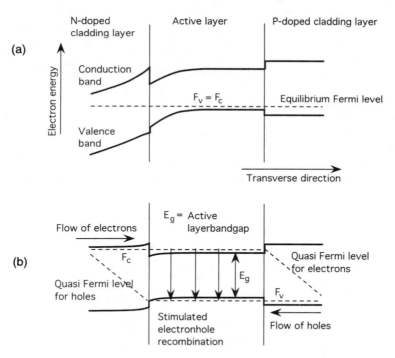

Figure 1.3 Energy band diagram of a DH laser diode with (a) zero bias and (b) forward bias $(F_C - F_V > E_g)$.

desirable situation. With the DH structure, a very effective carrier confinement is obtained in very thin active layers (<0.2 μm), which are much smaller than the diffusion length of the carriers (which is a few micrometers). In Chapter 5, a closer look will be taken at carrier injection and leakage phenomena.

In practice, more than three layers are used to obtain better electrical (e.g., contact resistance), optical (e.g., separate confinement techniques), or crystallographic (e.g., use of buffer layers) properties for the laser diodes.

Some limitations have to be taken into account when choosing semiconductor materials for the laser devices. The lattice constants of the different layers should only have a minimal mismatch. Otherwise the mechanical stress and the dislocation densities in the crystal become too large, especially at the heterojunction interfaces. This may lead to early damage and aging of the laser diode. Furthermore, one must keep in mind that direct semiconductors with smaller bandgaps do not always have larger refractive indexes [14].

The DH structure is usually made up of III-V compound semiconductors. In some cases, the II-VI system is also used. GaAs/Al$_x$Ga$_{1-x}$As laser diodes are used for the first fiber transmission window at around 0.8 μm, with typical values for x in the $0.25 < x < 0.65$ range. For the second window, at 1.3 μm, the active layer consists of the In$_{1-x}$Ga$_x$As$_{1-y}$P$_y$ compound semiconductor with $x = 0.28$ and $y = 0.6$ for lattice matching with InP cladding layers. For the third window, at 1.55 μm, the active layer also consists of the In$_{1-x}$Ga$_x$As$_{1-y}$P$_y$ compound semiconductor, but now with $x = 0.42$ and $y = 0.1$ for lattice matching with InP cladding layers. A more detailed account of the possibilities for combining active layer materials with cladding layer materials is given in [6] and [15], respectively, for AlGaAs and InGaAsP compounds.

Moreover, layers can be grown that are much thinner than the free carrier scattering distance. Depending on the thickness and the number of such ultrathin layers, one speaks of MQWs with only a few atom layers or super lattices with many layers, each only a few atoms wide. Devices with such layers show specific properties due to the quantum confinement of the carriers. In the next chapter, the gain in MQW active layers will be discussed further in some detail.

Confinement is not only a design target in the transverse direction. In the lateral direction, photon and carrier confinement is desired to obtain lateral single-mode behavior and lower threshold currents. Two mechanisms for lateral optical confinement are distinguished: gain-guiding and index-guiding.

In gain-guided lasers, the lateral confinement of the optical field is obtained by confining the area of gain to a certain lateral extent. This is realized by injecting the pump current into a contact stripe of limited width (typically 2 to 5 μm). Underneath, the stripe gain is created through carrier injection. The injected current will spread laterally to a certain extent, but at some distance away from the center of the stripe, the injected current density (A/cm2) will be so low that absorption will prevail in the

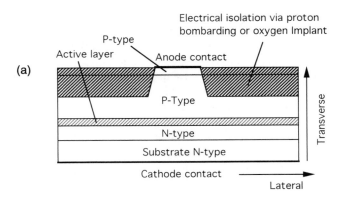
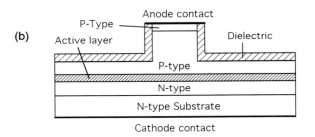
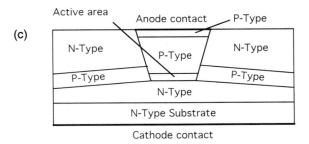

Figure 1.4 Laser diodes with different lateral optical guiding mechanisms: (a) gain guided laser, (b) ridge waveguide laser with weak index guiding, and (c) buried heterostructure laser with strong index guiding.

active layer. Therefore, the optical field will also be confined to the area underneath the stripe [16]. Figure 1.4(a) shows a proton-bombarded DH stripe laser. The proton-bombarded area limits the lateral spread of the current because the bombarded part is insulating.

If a lateral variation in the refractive index is introduced, lateral index-guiding can also be obtained. In lasers with a weak lateral index variation, combined index and gain-guiding will appear. Figure 1.4(b) shows a ridge waveguide DH stripe laser. The ridge structure causes a lateral variation of the effective refractive index [17], but this index variation is too weak to dominate the gain-guiding. Therefore, lateral optical confinement is realized by the cooperation of the two guiding effects.

When a strong lateral index variation is introduced, the index-guiding can become so strong that it entirely dominates the lateral waveguiding. This is the case in *buried heterostructure* (BH) lasers. Figure 1.4(c) illustrates a BH laser diode. In the BH, strong lateral and transverse index-guiding is coupled with a strong lateral and transverse carrier confinement. Several variants of the BH laser structure exist and the most important ones will be discussed in Chapter 5.

Notice that the more lateral index-guiding is desired in a laser diode, the more complex its fabrication usually becomes. On the other hand, BH lasers have a much more stable lateral field pattern and show lower threshold currents. Gain-guided and partially index-guided lasers have a tendency to show multiple lateral modes at higher power levels, and this results in multiple spectral lines [18].

As already mentioned, in a conventional Fabry-Perot laser, a cleaved facet mirror is used to obtain the optical feedback needed for laser operation. In DFB lasers, a corrugation, usually called *grating*, is introduced in one of the cladding layers, and the Bragg reflections at this periodic structure cause a very wavelength-sensitive feedback in the laser. Figure 1.5 depicts the general structure of a DFB laser. Stripe lasers, as well as ridge waveguide stripe lasers or BH laser diodes, can be fabricated with a grating. However, DFB lasers will most often be based on BHs because they show much better lateral single-mode and threshold behavior.

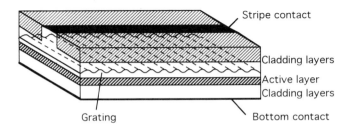

Figure 1.5 Example of a stripe contact DFB laser diode.

1.3 OPERATION OF THE LASER DIODE

1.3.1 The Basic Concept of Fabry-Perot Laser Diodes

The operation of a traditional Fabry-Perot laser diode is explained in a simplified way from a control system point of view. To this end, the basic Fabry-Perot scheme of Figure 1.2 is converted into the system block diagram of Figure 1.6. In this conversion, the spectral domain is introduced.

Figure 1.6 can be interpreted as follows. Assume an optical wave or mode, represented by the amplitude $A \exp(j\omega t)$, that starts at the left facet and travels through the laser toward the other facet; ω is the angular optical frequency. As the wave passes through the cavity, it is amplified with a factor $\exp(g_{mod}L/2)$. Here, g_{mod} is the modal gain that the wave experiences while traveling over a unit of length through the amplifying waveguide that constitutes the cavity. In other words, g_{mod} expresses the amplification of the intensity of the mode per unit of propagated distance. L is the cavity length. At the same time, the wave also receives a phase shift $\exp(-j\beta L)$, with β being the propagation constant of the modal field of the waveguide. At the right facet, part of the wave is reflected back into the cavity with amplitude reflection coefficient r_2 and part is transmitted. The transmitted wave is the output light of the laser. The reflected wave travels back in the cavity and it is again amplified with a factor $\exp(g_{mod}L/2)$ while it receives a further phase shift $\exp(-j\beta L)$. After reflection at the left facet, the amplitude of the wave has become $r_1 r_2 A \exp(g_{mod}L) \exp(-2j\beta L) \exp(j\omega t)$.

Figure 1.6 represents a positive feedback system with a saturable amplifier with gain $\exp(g_{mod}L)$. The feedback loop contains a feedback loss $r_1 r_2$ and a frequency filter with phase shift $\exp(-2j\beta L)$. The latter can also be expressed as $\exp(-2j\omega L n_e/c)$, with c the speed of light (in free space) and n_e the effective refractive index of the mode that travels through the laser waveguide.

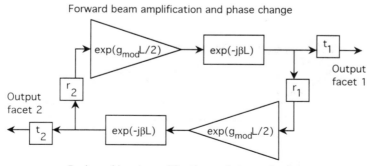

Figure 1.6 System view of a Fabry-Perot laser.

Introduction to Fabry-Perot and Distributed Feedback Laser Diodes

The feedback system of Figure 1.6 fulfills the condition for oscillation if

$$r_1 r_2 \exp(g_{mod}L) \exp(-2j\beta L) = 1 \quad (1.1)$$

This is equivalent to the following two resonance conditions

$$\omega = m\pi c/(Ln_e) \quad (1.2)$$

$$g_{mod} = (1/L) \ln(1/|r_1 r_2|) \quad (1.3)$$

with m being an integer.

The first condition is the phase resonance condition. It determines the possible wavelengths at which lasing can occur. These are called the *longitudinal modes* of the laser. Equation (1.2) implies that the laser can oscillate at an infinite number of discrete wavelengths, each separated by $\pi c/(Ln_e)$. In practice, lasing will not happen at all these wavelengths because the second condition, the amplitude resonance condition, requires the round-trip gain $|r_1 r_2| \exp(g_{mod}L)$ to equal 1, and because the optical gain in a direct semiconductor has an optical frequency dependence. Therefore, laser action is only possible in a limited wavelength range.

The amplitude of the generated light oscillation can be determined from the second condition. This condition expresses how much effective gain is needed to compensate for the mirror losses. In a laser diode, the required gain corresponds to a required carrier concentration in the active layer, and this in turn can be translated into the required injection current I. To determine the amplitude of the optical signal, one must take the nonlinear behavior of the amplifier into account. As long as there is no reasonable amount of optical power in the laser cavity, the modal gain g_{mod} will increase almost linearly with the injected current. However, at a certain current level, called the threshold current I_{th}, g_{mod} will be sufficient for the round-trip gain $\exp(g_{mod}L)$ to overcome the feedback losses. The stimulated emission, triggered by the spontaneous emission from spontaneous radiative recombinations, will then be able to build up an optical field for those wavelengths that fulfill the phase resonance condition and receive sufficient gain.

Through stimulated recombination, the optical field will prevent the carrier concentration from increasing when the current injection is further increased above the threshold level. Instead, an optical field will build up with an average optical power proportional to $I - I_{th}$. Figure 1.7 illustrates the optical power buildup as I passes through its threshold. In the same picture the gain clamping is also illustrated.

Because of the gain clamping and because the material gain is wavelength-dependent, the amplitude resonance condition will only be fulfilled for the resonant wavelength of (1.2) that has the highest modal gain. This implies that the laser will operate at only one wavelength, because the other resonant wavelengths on both sides

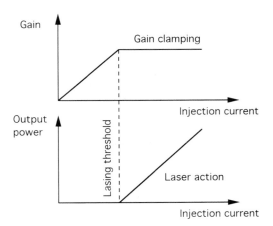

Figure 1.7 Optical power buildup and the associated gain clamping.

of the lasing wavelength do not receive enough gain to overcome the feedback losses. These longitudinal modes are called the *side modes*, and ideally they should be completely suppressed. However, in a Fabry-Perot laser diode, several side modes exhibit a gain that is only slightly smaller than the threshold gain. Due to the amplification of spontaneous emission that couples into those modes, they will show some power. Moreover, when the laser is modulated, modulation phenomena may raise the instantaneous gain above the threshold values of some of the side modes. In that case, the side modes will also start to lase and the laser will show a multimode spectrum under modulation.

1.3.2 Optical Material Parameters

In the above section, the modal gain g_{mod} (cm^{-1}) refers to the gain minus the losses experienced by the basic transverse/lateral mode that propagates back and forth through the active waveguide that forms the laser diode. This gain is the effective gain of the modal field of the waveguide. It is related to the local material gain (g) in the active layer and the material losses (α) in the different layers via some weighted average (see Chapter 3). Similarly, n_e is the effective refractive index of the propagating transverse/lateral mode. It is a weighted average of the real refractive indexes of the different layer materials that make up the waveguide. In practice, g_{mod} and n_e will be dependent on the axial position along the laser.

Material gain and absorption can be seen as imaginary parts of the refractive index. Under carrier injection, the real part of the refractive index will also be influenced. Therefore, the local complex refractive index in a layer can in general be written as

$$n_c(\lambda) = n_{c0}(\lambda) + \Delta n_g(\lambda) + jg(\lambda)/(2k_0) + \Delta n_\alpha(\lambda) - j\alpha(\lambda)/(2k_0) \tag{1.4}$$

Here $n_{c0}(\lambda)$ is the complex refractive index without carrier injection, g is the local power gain of the material (with $k_0 = 2\pi/\lambda$), and $\Delta n_g(\lambda)$ is the associated variation of the real refractive index part. The gain g and Δn_g are due to stimulated radiative transitions between the conduction and valence bands. The absorption caused by other interband transitions and also by intraband effects is represented by $\alpha(\lambda)$ [19]. The variation of the real refractive index corresponding with α is Δn_α.

Because g is strongly dependent on the carrier density, it is clear that modulation of the carrier density through modulation of the injected current will cause an intensity modulation of the optical output power. At the same time, the refractive index Δn_g will also be modulated, causing chirp or frequency modulation.

1.3.3 Thermal Aspects of Laser Diodes

Traditionally, the temperature dependence of the threshold current of DH lasers is described by $I_{th}(T) = I_0 \exp(T/T_0)$, where I_0 and T_0 are the characteristic current and temperature, respectively, of the laser. For AlGaAs lasers, $T_0 > 120K$ holds near room temperature, while for InGaAsP lasers, T_0 lies in the 50K to 70K range. A lower value of T_0 implies that the threshold current is more temperature-sensitive. Recently, a more accurate expression for the temperature dependence of the threshold current density J_{th} has been proposed [20]

$$J_{th} = \left[\frac{p}{C}(T_{max} - T)\right]^{\frac{1}{p}} \tag{1.5}$$

where C is an essentially temperature-independent constant that depends on the active layer structure, p is approximately 1/2, and T_{max} is a characteristic temperature.

At ambient room temperature, the maximum output power of a laser may be limited by thermal runaway. In such a runaway process, more and more injection current is required to offset the effect of the internal temperature increase with higher currents and power, and this in turn further increases the internal temperature.

Thermal effects can be due to structure-related effects or to more fundamental effects. Structure-related effects occur when part of the injected current can flow around the active region. This leakage current can vary in magnitude with temperature and result in abnormal temperature sensitivities of the laser diode characteristics. Facet heating (and the associated facet degradation) is another structure-related effect. It is caused by high output powers, together with surface recombinations and lower heat transfer capabilities at the facets.

More fundamental temperature sensitivities are related to carrier leakage across the DH heterojunction and to the material properties of the active layer material.

DH heterojunction carrier leakage can be large for InGaAsP-InP lasers with emission wavelengths below 1.1 µm because of their small heterojunction barrier height. For lasers in the typical optical communications wavelength ranges around 1.3 and 1.55 µm, the barrier heights are larger and drift and diffusion modeling of the carrier leakage across the heterojunction shows limited leakage. Heterojunction carrier leakage can also be strongly reduced by proper design choices of the doping and thickness of the cladding layers.

Several material properties that are temperature-dependent and that influence the behavior of the laser can be distinguished: nonradiative recombination such as Auger recombination, absorption phenomena, and the stimulated emission resulting in the gain. Also, the optical losses in the cladding layers may be temperature-dependent. Auger recombination processes have a considerable influence on the thermal behavior of InGaAsP lasers [15].

The temperature dependence of the gain has its origin in two phenomena. The bandgap decreases with rising temperature and the Fermi-Dirac distributions of the electrons and the holes become spread over larger energy ranges. The first phenomenon results in a shift of the gain peak wavelength. The temperature coefficient of the gain peak wavelength is around 0.5 nm/K. The second phenomenon causes a decrease of the gain with temperature [6]. Therefore, at higher temperatures, more pumping power is required to maintain the same level of gain [6].

The change of the refractive index with temperature is smaller than the corresponding change in gain. In a laser operating at threshold, the refractive index will change with temperature due to its direct temperature dependence, but also because the carrier density in the active layer increases with temperature. The latter increase corresponds to the need for more pumping with higher temperatures. The change in real refractive index with temperature causes the resonant lasing wavelength to shift a little. However, because the temperature coefficient of the gain peak wavelength is much larger than that of the resonant lasing wavelength, the lasing mode in Fabry-Perot lasers jumps from one longitudinal mode to another when the temperature is raised over only a few degrees. On the other hand, DFB lasers operate without mode jumps over much larger temperature ranges because of the frequency selectivity of the optical losses induced by the grating in the DFB laser diode.

In practice, many lasers are mounted on a Peltier element to stabilize their operating temperature. Very often the Peltier element is mounted inside the laser package.

By adapting the current through the Peltier element, the temperature of the laser can be changed and the operating wavelength can be tuned. This thermal tuning allows calibration of the laser wavelength. The dynamics of the thermal effects are only relevant at low modulation frequencies below about 100 MHz. In the remainder of this book, thermal effects will not be discussed in detail, because these effects have little impact when the lasers are properly used.

1.4 ESSENTIAL LASER DIODE CHARACTERISTICS

In optical communication systems, the laser diode is used as a signal transmitter, signal amplifier, wavelength reference, or wavelength-tunable local oscillator. Depending on the type of application, the external laser characteristics have to comply with certain minimum specifications. Below, the most important laser diode characteristics are discussed. They are grouped into static and dynamic characteristics depending on the type of laser excitation. The figures in this section relate to a simple Fabry-Perot laser diode unless explicitly specified. The case of the DFB laser will be elaborated in subsequent chapters.

1.4.1 Static Characteristics

The following static characteristics are distinguished: P/I curve, spectral properties, and noise characteristics such as linewidth and *relative intensity noise* (RIN).

P/I Curve

Figure 1.8 shows a P/I curve in which the total output power at one of the facets of the laser diode is depicted as a function of the drive current. The threshold current is

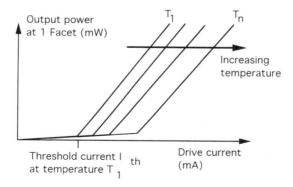

Figure 1.8 P/I characteristics.

indicated in the figure. Low threshold currents are preferred because they simplify the driver electronics and cause less thermal stress on the laser, increasing the laser lifetime. To minimize the threshold current, BH structures with very low leakage currents and high-gain MQW active layers have to be used. Furthermore, all internal photon and carrier losses have to be reduced. Therefore, good-quality epitaxial layers and heterojunction interfaces are important in order to reduce nonradiative carrier recombination. This also reduces the optical scattering losses due to imperfections of the layer interfaces. Diffraction losses can be limited through strong index-guiding. Lowering the internal losses also favors a high external efficiency $\eta_{ext} = dP/dI$.

A linear P/I curve over a large current range allows for simple control of the optical power and for a larger modulation depth. However, this also requires that the leakage currents and the temperature do not increase with the drive current I.

Spectral Properties

The spectral distribution of the optical power is an important characteristic for communications systems. Figure 1.9 shows the optical spectrum. In general, several spectral lines may be present, corresponding to multiple longitudinal and lateral modes. Usually the dominant spectral line is called the *lasing mode*, while the other lines are referred to as the *side modes*. Optical communication systems require that, even under modulation, only one mode is present. DSM lasers are devices that meet this requirement. Typically, laser diodes are called DSM lasers if their side modes are suppressed by 30 dB or more with respect to the lasing mode.

In heterodyne detection systems, laser diodes can be used as local oscillators. In this case, the lasing mode must have a tunable wavelength because the local

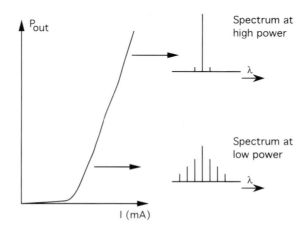

Figure 1.9 The optical spectrum of a Fabry-Perot laser diode at two different output power levels.

oscillator laser must be tuned to the received optical carrier. It is preferable for the wavelength to be continuously tuned, without mode jumps. This simplifies the wavelength control of tunable devices. Special multisection DFB and DBR lasers (see Section 1.6 for further details on DBR lasers) can achieve these requirements over limited wavelength ranges (3 to 10 nm) [21]. The essential characteristic here is the λ/I curve. Wavelength tuning based on changing bias currents is possible because the refractive index in the laser cavity also varies with the carrier density in the active layer. To fine-tune the wavelength of DFB lasers that have to operate at a fixed wavelength, the temperature is usually varied. In this case, wavelength tuning is based on the temperature dependence of the refractive index of the semiconductors that constitute the laser waveguide and especially the active layer.

Noise Properties

The optical output of a laser exhibits intensity noise and phase noise. These are caused by fluctuations in the photon and carrier populations in the laser. Relevant fluctuation phenomena include the spontaneous emission fluctuations and the carrier shot noise.

The optical phase noise basically determines the linewidth. This is the width of the power spectrum of the lasing mode at half of its maximum intensity. Its value typically falls in the 1- to 100-MHz range. At low and moderate power levels, the linewidth decreases with increasing power, but at higher powers, linewidth saturation or linewidth rebroadening may occur. In coherent receivers for optical communications systems, narrow linewidths lead to substantial gains in the receiver sensitivity.

In analog or digital amplitude modulated systems, a small intensity noise is desired. Intensity noise (i.e., the fluctuations in the emitted optical power) is characterized by the RIN. The RIN of a mode is defined as the spectral density of the intensity noise of that mode relative to the square of the average photon density of the mode. Integrated over a certain frequency window Δf, this is the RIN one would measure with a photodetector with bandwidth Δf.

1.4.2 Dynamic Characteristics

The dynamic characteristics can be divided into large-signal and small-signal properties.

Large-Signal Properties

Large-signal IM characteristics include the turn-on delay, rise time (10% to 90%), overshoot, damping, and the oscillation frequency of the transient response of the optical output power to a step input current. The large-signal response of a Fabry-Perot laser is illustrated in Figure 1.10, which shows the response of the optical power

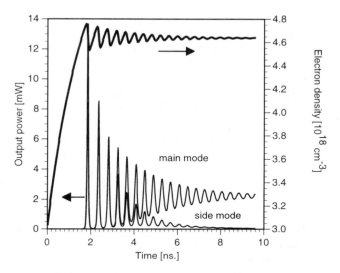

Figure 1.10 The response of a laser to a large-signal step current of finite duration.

in main and side modes and the response of the electron density to a step current. The observed large-signal properties depend on the current levels before and after the step because of internal nonlinear effects. For a non-DSM laser, the total optical power is divided over different longitudinal modes during a transient response, as can be seen in Figure 1.10. The transient oscillation causes the gain of the different modes to vary around their static values. Therefore, the gain of a side mode can be temporarily larger than its losses, which will temporarily excite the side mode.

A refractive index variation is associated with the gain oscillations that causes transient behavior of the wavelength of the lasing mode(s). This behavior is called *chirp* when it is considered undesirable, as in the case for IM systems. In IM optical fiber communications systems, the chirp and the excited side modes can cause dispersion problems. On the other hand, in FM-based systems, the wavelength variation of the lasing mode is used in a constructive way. However, in this case, the IM effects are usually considered undesirable side-effects.

In digital IM systems, pattern effects may also occur for high modulation speeds. This means that the pulse pattern of the outgoing optical power will show more or fewer pulses than the input current pulse sequence, leading to bit errors. The reason behind this is a too small damping and a too large delay. A proper choice of the upper and lower current levels of the modulation current sometimes allows the suppression of this phenomenon without having to lower the bit rate.

Small-Signal Properties

The most important small-signal characteristics are the IM and FM responses. These modulations appear if the injection current is modulated around its dc bias point with a small-signal current with modulation frequency $\Omega/2\pi$. Thermal effects of a dynamic nature only occur at low speeds (<100 MHz) and will be neglected.

The first-order small-signal FM and IM characteristics are a linear approximation of the dynamic behavior of laser diodes. The nonlinearity of the laser diode, however, results in an output power that is not perfectly linear as a function of the injected modulation current, and this nonlinearity can be represented by the higher order harmonic and intermodulation distortions. The harmonic distortion describes the output at the higher order harmonics of the modulation frequency, while the intermodulation distortion describes the output at the sum and difference frequencies when the current modulation contains several frequencies. This book will mainly focus on second- and third-order harmonic distortion of the output power.

1.5 USE OF LASER DIODES IN OPTICAL COMMUNICATIONS SYSTEMS

Figures 1.11 and 1.12 show two basic types of optical communications systems: direct detection systems and heterodyne/homodyne detection systems. In both types of systems the transmitter consists of a laser diode that is either modulated directly (via the injection current) or externally (via an optical modulator). The modulation can take many forms, including amplitude, phase, or frequency modulation.

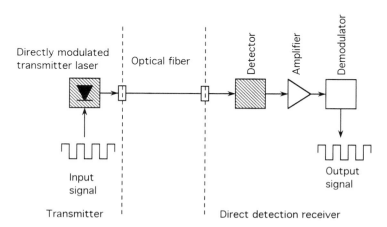

Figure 1.11 Fiber-optic transmission system with a direct detection receiver.

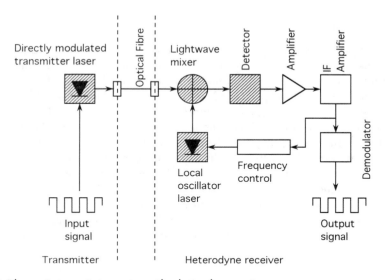

Figure 1.12 Fiber-optic transmission system with a heterodyne receiver.

In a direct detection system, as shown in Figure 1.11, the received optical signal is directly converted in an electric current by means of a photodiode. The current generated by the photodiode is proportional to the received power. Several analog and digital modulation schemes are used in direct detection systems. The choice of the modulation to be used for a certain application depends on cost, complexity, and the signal-to-noise ratio (SNR) improvement versus bandwidth occupancy trade-off.

Optical communication schemes employing *subcarrier multiplexing* (SCM) with analog modulation of the optical carrier are applied in CATV distribution networks. High-speed analog modulation is also used in wireless telephony remote antenna feeders. The performance of such analog systems depends strongly on the linearity of the system. Nonlinearities, such as the distortion products, degrade the sensitivity of the system, resulting in a reduced *carrier-to-noise ratio* (CNR) limit for the transmission link. In a fiber transmission system, the detector and receiver are generally very linear, and therefore the level of distortion depends mainly on the linearity of the transmitter laser. In CATV systems the harmonic distortions are usually required to be below −65 dBc (i.e., magnitude of the distortions in decibels below the unmodulated carrier). DFB lasers are the desired transmitters for such systems, since they usually also have a small intensity noise.

In a homodyne/heterodyne detection receiver, the received light signal will first be mixed with the optical signal from the local laser in the receiver. If the optical frequencies ω_l and ω_r of the local laser and the received signal, respectively, are equal, one talks about a homodyne receiver. In the other case, the receiver is a heterodyne receiver. For the heterodyne/homodyne receiver, the current generated by the photodiode is proportional to the square root of the product of the optical power of the local laser output and the received light signal. This product lies in the baseband for homodyne receivers or around the intermediate frequency ω_l-ω_r for heterodyne receivers. Heterodyning or homodyning considerably improves the SNR. Depending on the used digital modulation schemes, further improvements in sensitivity are possible.

Essential to the performance of heterodyne solutions is the spectral purity of the laser diodes, especially for coherent demodulation schemes. Moreover, in multiwavelength systems, the heterodyne receiver can operate as a wavelength channel selector if a tunable local laser oscillator is used. For this purpose, multielectrode DFB lasers have been developed.

1.6 DYNAMIC SINGLE-MODE LASER DIODES

Even if a laser diode shows single-mode behavior under CW operation, it may show a multimode spectrum under modulation. Several laser structures have been designed for achieving single-mode operation under modulation conditions. DSM operation is achieved if the threshold gain of the lasing mode is significantly smaller than the threshold gain of the other longitudinal modes.

A laser diode is usually considered to be a DSM laser if the ratio of the power in the lasing mode P_L to the power in any of the side modes P_S is larger than 30 dB [22].

DSM operation can be obtained by increasing the internal cavity losses or feedback losses of the side modes. This requires that these losses are made much more wavelength-dependent.

Another way of obtaining DSM operation is by decreasing the material gain experienced by the side modes. Using the inherent wavelength dependence of the material gain, DSM operation will imply a larger longitudinal mode spacing.

Several laser diode architectures have been developed to meet the DSM objective. The main ones are short lasers, coupled-cavity lasers, injection-locked lasers, and lasers with distributed feedback. Figure 1.13 summarizes these different DSM laser structures. A short description is given below.

1.6.1 Short-Cavity Lasers

Short laser cavities cause the wavelength spacing between the adjacent longitudinal modes to be large. If the spacing is large enough, only one mode will have a wavelength near the gain peak and DSM operation can be expected. However, very short lasers are needed, and this implies high gain or very high facet reflectivities, leading to low external efficiencies. In practice, this approach does not deliver acceptable results.

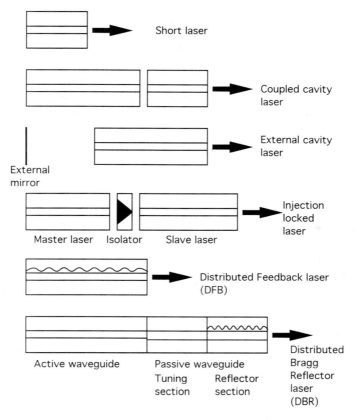

Figure 1.13 Axial side views of different types of DSM laser structures. (*After:* T. E. Bell, *IEEE Spectrum*, December 1983, p. 38.)

1.6.2 Coupled-Cavity Lasers

Figure 1.13 shows two types of coupled-cavity lasers: external-cavity lasers and C^3 lasers. In these devices, the light propagating through the cavity experiences multiple reflections that all interfere with each other. The result of this interference is very wavelength-sensitive feedback. In this way, closely spaced longitudinal modes that experience too small of a difference in material gain can have a large difference in feedback losses, leading to DSM operation.

The disadvantage of these devices is their high sensitivity to variations in cavity lengths, temperature changes, and even to modulation depths that are too strong. In the latter case, mode jumps may occur when the laser is modulated.

In practice, only simple external-cavity lasers have found some applications, mainly in environments with very stable conditions, such as laboratory setups. Long external-cavity lasers offer narrow linewidths. Wavelength tuning can be realized with them by changing the cavity length or by changing the reflectivity of the external mirror in case the external mirror is frequency-selective. Calibration of external-cavity length, temperature, and axial alignment is usually required.

1.6.3 Injection-Locked Lasers

The laser for which DSM operation is desired, here called the slave laser, is frequency-locked to a master laser. The master laser is a laser operating at a low power level and with a single mode. The master laser can be an index-guided Fabry-Perot laser operating under a static drive current. Injection locking will only occur if the lasing wavelength of the slave laser in stand-alone operation is sufficiently close to the wavelength of the master laser. When locking of the slave wavelength to the master wavelength occurs, the slave laser can be modulated with DSM behavior.

To keep the master laser stabilized to its operating point, no light of the slave laser may couple into the master laser. Therefore, an isolator must be used between the master and the slave. This makes this type of setup rather elaborate. Therefore, it is mainly used in laboratory setups.

1.6.4 Laser Diodes With Distributed Optical Feedback

Two important types of lasers can be distinguished in this class of DSM lasers: DBR and DFB laser diodes. Both types are illustrated in Figure 1.13.

In simple Fabry-Perot lasers, a cleaved facet mirror is used to obtain the feedback for laser oscillation. In DBR and DFB lasers, the optical feedback is obtained by the periodic variation of the effective refractive index of a corrugated optical waveguide inside the laser cavity. The periodic index variation causes a very wavelength-selective feedback. Maximum reflectivity of the grating occurs around the Bragg wavelength λ_B of the grating. The Bragg wavelength is given by

$$\lambda_B = 2\Lambda n_e \quad (1.6)$$

where Λ is the period of the grating and n_e is the effective refractive index of the waveguide without grating. Figure 1.14 shows the reflection characteristics of a passive lossless corrugated waveguide of length L as a function of detuning $\lambda - \lambda_B$. This clearly illustrates the wavelength-sensitive reflection properties of Bragg gratings.

To explain this reflection characteristic in an intuitive way, reference is made to Figure 1.15, where an infinite waveguide with a rectangular index grating of finite length is illustrated. In such a waveguide, the effective refractive index will vary as shown in the same figure. Consider now a wave that is incident from the left on the corrugated section. At each of the index jumps, part of this wave will be reflected and

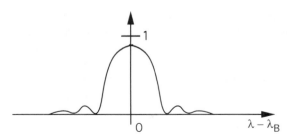

Figure 1.14 The reflection characteristic of a corrugated section of length L as a function of detuning $\lambda - \lambda_B$.

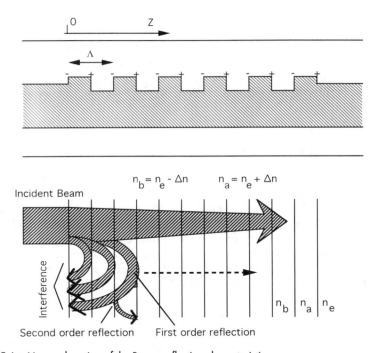

Figure 1.15 Intuitive explanation of the Bragg reflection characteristic.

will travel backwards. All of these reflected waves will only partially reach the starting point of the grating (chosen to have axial coordinate $z = 0$) because part of them will in turn be reflected toward the left at the index jumps. Notice, however, that these second-level reflections will also undergo further reflections. Finally, some part of each of the reflected waves will pass through $z = 0$ and add up to represent the reflection of the incident wave at $z = 0$. Depending on the wavelength λ of the incident light, the reflected waves will interfere constructively or destructively at $z = 0$. For the Bragg wavelength λ_B, the light experiences a phase shift π when traveling over a distance Λ. Furthermore, the phases of the subsequent reflectivities along the grating alternate between 0 and π. Therefore, when $\lambda = \lambda_B$ all the interfering reflections add up constructively when they arrive at $z = 0$. Consequently, at $z = 0$, the incident wave will be strongly reflected for the Bragg wavelength λ_B. For other wavelengths around λ_B, all those interfering reflections will suffer from different phase shifts leading to a lower overall reflection coefficient at $z = 0$ for the incident wave. This clarifies the wavelength sensitivity of the reflection characteristic of Figure 1.14. This sensitivity is further influenced by the length of the grating and the size of the periodic index variation.

In DBR lasers, the grating is fabricated in a waveguide external to the active region. This external waveguide must show few optical losses in order to obtain good wavelength selectivity and strong reflectivity from the grating. Moreover, the coupling efficiency between the active and passive waveguide structures in a DBR laser should be high in order to gain high-performance operation. The external waveguide has the advantage that it can filter the lateral TM mode out of the laser light.

In DFB lasers the grating is constructed in the active waveguide itself. Therefore, DFB lasers are simpler to fabricate. However, knowing how to integrate active and passive waveguides in the same monolithic component with high optical coupling efficiency, as is required for DBR lasers, is an important technological step in the development of *photonic integrated circuits* (PIC) or *optoelectronic integrated circuits* (OEIC). DBR lasers are more complicated to fabricate, but on the other hand they are easier to control. If DBR lasers are equipped with a phase control tuning section (see Figure 1.13), wavelength tuning can be achieved in a quite straightforward way. Therefore, DBR lasers are popular devices as wavelength-tunable laser diodes. However, DFB lasers also lend themselves to wavelength tuning [21].

Up to now, we implicitly assumed DFB lasers in which the grating only creates a periodic variation in the real part of the refractive index. Such lasers are called *index-coupled DFB lasers*. At the time of publication, by far most of the fabricated DFB lasers are of this type. DFB lasers may, however, also be constructed with periodic gain or loss variations [23]. Such lasers are called *gain-coupled DFB lasers*. The explanation related to Figure 1.15 remains valid for gain coupling. The main difference is that the phases of subsequent reflectivities along the gain/loss grating now alternate between $-\pi/2$ and $\pi/2$.

In practice, partly gain-coupled DFB lasers will be more common than purely gain-coupled DFB lasers, because in realistic device structures, gain or loss variations are usually accompanied by larger real refractive index variations. Lasers with both index and gain coupling are referred to as *complex-coupled DFB lasers*. Even with only a small and partial gain coupling, such complex-coupled DFB lasers show an outstanding performance as far as single-mode yield is concerned (see Chapter 6). Therefore, these devices are rapidly expanding their commercial presence. This book covers both index- and gain-coupled DFB lasers.

1.7 ORGANIZATION OF THIS BOOK

The aim of this book is to give the reader a thorough understanding of the operation of DFB lasers. To this end, the material in this book has been roughly divided into three parts. The first part consists of Chapters 2 to 5 and covers the theory. The second part corresponds to Chapters 6 to 9 and covers the external characteristics of DFB laser diodes. Finally, the last part consists of Chapter 10, which deals with fabrication techniques.

Chapters 2 to 5 describe the fundamental theoretical concepts of laser diodes. In Chapter 2, a general rate equation model is presented. This model is valid for Fabry-Perot lasers as well as DFB lasers. However, for DFB lasers, some additional simplifying assumptions have to be made.

A more accurate DFB laser model is given in Chapter 3, where the coupled-mode theory of DFB lasers is described. The coupled-mode theory outlined there takes into account multiple longitudinal modes, axial variations, dynamic properties, and noise characteristics. In this way the theory given in this book differs considerably from the usual textbook material. However, the level of detail used in this book is essential to come to a proper understanding of the detailed device behavior of DFB lasers.

In Chapter 4, the theoretical framework as developed in Chapter 3 is further explained and its application is illustrated with some classical examples. Also, the link between the simpler rate equation model and the more complete coupled-mode model is clarified.

Chapter 5 concentrates on carrier injection phenomena in laser diodes. In particular, carrier leakage and parasitic effects due to the heterojunction structure of the laser will be discussed. The efficiency and the dynamics of carrier injection can have a detrimental impact on the output properties of a laser diode. Much of the material discussed in this chapter holds for laser diodes in general and is not limited to DFB lasers.

Next, Chapters 6 to 9 discuss the properties of DFB lasers in detail. Each of these chapters discusses a different set of important laser characteristics. Chapter 6 deals with the spectrum of DFB lasers. Chapter 7 discusses the linear dynamic behavior of DFB lasers with emphasis on the FM and IM response. Chapter 8 handles the nonlinear harmonic distortion properties of laser diodes, while Chapter 9 covers the noise properties of DFB laser diodes.

Finally, Chapter 10 explains the basic DFB laser diode fabrication and packaging techniques, while the epilogue provides some concluding remarks.

References

[1] Oda, K., M. Fukui, M. Fukutoku, A. Umeda, T. Kitoh, and H. Toba, "10 Channel × 10 Gb/s Over 500 km Optical FDM-Add/Drop Multiplexing Experiment Employing a 16-Channel Arrayed-Waveguide-Grating ADM Filter," *Proc. ECOC 95*, Brussels, 17–21 September 1995, pp. 59–62.

[2] Motoshima, K., K. Takano, J. Nakagawa, and T. Kitayama, "Eight-Channel 2.5 Gbit/s WDM Transmission Over 275 km Using Directly Modulated 1.55 μm MQW DFB-LDs," *Technical Digest OFC'95*, San Diego, 26 February–3 March 1995, pp. 120–121.

[3] King, J. P., I. Hardcastle, H. J. Harvey, P. D. Greene, B. J. Shaw, M. G. Jones, D. J. Forbes, and M. C. Wright, "Polarisation Independent 20 Gbit/s Soliton Data Transmission Over 12500 km Using Amplitude and Phase Modulation Soliton Transmission Control," *Proc. ECOC 95*, Brussels, 17–21 September 1995, pp. 291–294.

[4] Mestdagh, D. J. G., *Fundamentals of Multiaccess Optical Fiber Systems*, Norwood, MA: Artech House, 1995.

[5] Lee, B. G., M. Kang, and J. Lee, *Broadband Telecommunications Technology*, Norwood, MA: Artech House, 1993.

[6] Casey, H. C., and M. B. Panish, *Heterostructure Lasers*, Parts A & B, New York: Academic Press, 1978.

[7] Kogelnik, H., and C. V. Shank, "Coupled-Wave Theory of Distributed Feedback Lasers," *J. Appl. Phys.*, Vol. 43, No. 5, May 1972, pp. 2327–2335.

[8] Nakamura, M., A. Yariv, H. W. Yen, S. Somekh, and H. L. Garvin, "Optically Pumped GaAs Surface Laser With Corrugation Feedback," *Appl. Phys. Lett.*, Vol. 22, 1973, p. 515.

[9] Aika, K., M. Nakamura, J. Umeda, A. Yariv, A. Katzir, and H. W. Yen, "GaAs-GaAlAs Distributed Feedback Diode Lasers With Separate Optical and Carrier Confinement," *Appl. Phys. Lett.*, Vol. 27, 1975, pp. 145–146.

[10] Yoshikuni, Y., K. Oe, G. Motosugi, and T. Matsuoka, "Broad Wavelength Tuning Under Single-Mode Oscillation With a Multi-Electrode Distributed Feedback Laser," *Electron. Lett.*, Vol. 22, 1986, pp. 1153–1154.

[11] Agrawal, G., J. Geusic, and P. Anthony, "Distributed Feedback Lasers With Multiple Phase-Shift Regions," *Appl. Phys. Lett.*, Vol. 53, July 1988, pp. 178–179.

[12] Thijs, P. J. A., L. F. Tiemeijer, J. J. M. Binsma, and T. Van Dongen, "Progress in Long-Wavelength Strained-Layer InGaAs(P) Quantum-Well Semiconductor Lasers and Amplifiers," *IEEE J. Quant. Electron.*, Vol. 30, 1994, pp. 477–499.

[13] Morthier, G., P. Vankwikelberge, K. David, and R. Baets, "Improved Performance of AR-Coated DFB Lasers by the Introduction of Gain Coupling," *IEEE Phot. Tech. Lett.*, Vol. 2, March 1990, pp. 170–172.

[14] Toda, M., "Material Selection for Double Heterojunction Lasers-A Higher Bandgap Does Not Necessarily Mean Lower Refractive Index," *IEEE J. Quant. Electron.*, Vol. QE-23, No. 5, May 1987, pp. 483–486.

[15] Agrawal, G. P., and N. K. Dutta, *Long-Wavelength Semiconductors Lasers*, New York: Van Nostrand Reinhold, 1986.

[16] Van de Capelle, J. P., P. Vankwikelberge, and R. Baets, "Lateral Current Spreading in DH-Lasers Above Threshold," *IEE Proc.*, Vol. 133, Pt. J, No. 2, April 1986, pp. 143–148.

[17] Thomson, G., *Physics of Semiconductor Lasers*, New York: Van Nostrand Reinhold, 1980.

[18] Streifer, W., R. Burnham, and D. Scifres, "Dependence of Longitudinal Mode Structure on Injected Carrier Diffusion in Diode Lasers," *IEEE J. Quant. Electron.*, Vol. QE-13, June 1977, pp. 403–404.

[19] Mozer, A., K. Romanek, O. Hildebrand, W. Schmid, and M. Pilkuhn, "Losses in GaInAs(P)/InP and GaAlSb(As)/GaSb Lasers—The Influence of the Split-off Valence Band," *IEEE J. Quant. Electron.*, Vol. QE-19, No. 6, June 1983, pp. 913–916.

[20] Evans, J. D., and J. G. Simons, "New Insight Into the Temperature Sensitivity of the Threshold Current of Long Wavelength Semiconductor Lasers," *Proc. of the IEEE Semiconductor Laser Conference*, Hawaii, 1994, pp. 237–238.

[21] Kuindersma, P. I., W. Scheepers, J. M. H. Cnoops, P. J. A. Thijs, G. L. A. Van Den Hofstad, T. Van Dongen, and J. J. Binsma, "Tunable Three-Section, Strained MQW, PA DFB's With Large Single Mode Tuning Range (72 Å) and Narrow Linewidth (around 1 MHz)," *Proc. 12th IEEE International Semiconductor Laser Conference*, Davos, Switzerland, 1990, pp. 248–249.

[22] Suematsu, Y., S. Arai, and F. Koyama, "Dynamic-Single-Mode Lasers," *Optica Acta*, Vol. 3, September–October 1985, pp. 1157–1173.

[23] Nakano, Y., Y. Luo, and K. Tada, "Facet Reflection Independent, Single Longitudinal Mode Oscillation in a GaAlAs/GaAs Distributed Feedback Laser Equipped With a Gain-Coupling Mechanism," *Appl. Phys. Lett.*, Vol. 55, 1989, pp. 1606–1608.

Chapter 2

Rate Equation Theory of Laser Diodes

The lumped rate equations constitute the simplest possible, but still accurate, mathematical description of laser diodes. They express the evolution in time of the average carrier density in the laser cavity, the average photon density in each cavity mode and the relative phase (or frequency) associated with each mode. They form a set of coupled first-order differential equations that are easily solved for a variety of driving conditions, and therefore allow one to gain much insight into the physics that govern the behavior of laser diodes. For this reason, the rate equations are still often used by laser experts having more advanced laser simulation programs at their disposal. The coupling of the different rate equations is caused by the carrier density dependence of the stimulated emission, the absorption, and the refractive index in an active material.

Since the derivation of the rate equations is relatively straightforward, the emphasis of this chapter is on the analytical formulas that can be obtained from a small-signal solution of the rate equations. Several of the important characteristics for telecommunications can be described explicitly using this approach. The resulting analytical expressions contain structural parameters as well as several material parameters expressing the dependence of gain, absorption, and refractive index on carrier density. This carrier density dependence is therefore discussed in more detail before the discussion of the analytical solutions. We conclude this chapter by introducing the complex topic of the influence of external reflections (originating, for example, from a lens, a fiber, a photodiode).

The rate equations and in particular their approximate analytical solutions will be used from time to time in Chapters 8, 9, and 10 as well. It can be noticed at this point that the rate equations for carrier and photon density as well as most of their solutions discussed in this chapter are valid for both Fabry-Perot and DFB lasers. Therefore, no distinction between these laser types is made in this chapter. An exact derivation of the rate equations for DFB lasers will be given in Chapter 4. It shows that only the phase equation becomes modified due to the distributed feedback, while quantities such as gain and cavity loss become more complex functions of power, wavelength, and carrier density than is assumed here.

2.1 CARRIER DENSITY RATE EQUATION

The carrier density rate equation expresses the conservation of electrons (or holes) inside the active layer. It follows from the continuity equations for the electron density and the hole density (well-known in semiconductor physics) [1]. If one assumes that all the injected current reaches the active layer (Figure 2.1) and that electron N and hole density P are uniform across the active layer, it can be expressed as

$$\frac{\partial N}{\partial t} = \frac{1}{qd}\left[J_n(d) - J_n(0)\right] + (G - U)_{av} \tag{2.1}$$

with $J_n(d)$ the electron current density at the border of the active layer and the N-type cladding, and $J_n(0)$ the electron current density at the border of the active layer and the P-type cladding. $(G - U)_{av}$ is the net rate of carrier generation and recombination averaged over the active layer. The electron current in the P-type cladding $J_n(0)$ can be neglected, since the larger bandgap of the cladding layers prevents electrons in the active layer from leaking into the cladding. The hole current in the N-type cladding can be neglected for the same reason, and hence $J_n(d) = J(d) = J$. Taking into account the exact form of the recombination rate, consisting of spontaneous and stimulated recombination, one finally arrives at

$$\frac{dN}{dt} = \frac{J}{qd} - AN - BN^2 - CN^3 - \sum_m G(N, \lambda_m)S_m \tag{2.2}$$

in which the summation represents the stimulated emission and is over all cavity modes (with mode index m). AN represents spontaneous carrier recombination via traps and at surfaces [2]. This recombination rate is proportional to the carrier density, since only one carrier (besides one trap) is involved in the process. BN^2 represents the bimolecular recombination (or spontaneous emission) rate [3], and this rate is proportional to N^2, since two carriers (an electron and a hole) are now involved. The CN^3 term stands for the Auger recombination rate [4]. In this process, the energy released during an electron-hole recombination is transferred to another electron (or hole), which is excited to a high energy state and relaxes back by losing energy to lattice vibrations. This Auger recombination, which is proportional to N^3 as a result of the three carriers involved, is generally the dominant nonradiative process in long-wavelength lasers (as used in advanced optical communication).

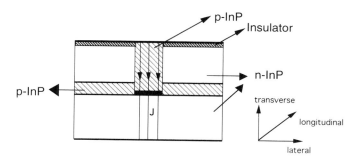

Figure 2.1 Current injection in a typical index-guided laser.

Typical values for A are in the range 1 to 5 ns^{-1}, typical values for B are in the range 0.5 to 2×10^{14} cm^3/s, and typical values for C are in the range 5 to 20×10^{19} cm^6/s.

The stimulated emission term in (2.2) will be discussed later, but we can already notice that the stimulated emission rate (total stimulated emission per unit time and unit volume in a mode) is proportional to the average photon density S_m in the mode and to a gain coefficient G (with unit s^{-1}). This gain coefficient depends on the carrier density and on the wavelength. It also depends a little on the photon densities in the different modes, so that in reality the stimulated emission rate saturates at high photon densities.

2.2 PHOTON DENSITY RATE EQUATION

The photon density rate equations express the conservation of photons in the different modes. The different processes that are involved now are stimulated and spontaneous emission (through which photons are created) and absorption and mirror loss (through which photons are lost). The stimulated emission rate and the loss rate are, at least in the first order, proportional to the average photon density. The proportionality between mirror loss and photon density is understood as follows. The photon flux that leaves a laser facet equals the photon density (at the facet and carried by the wave propagating toward the facet under consideration) times the group velocity. This photon density at the facet and carried by the wave propagating toward the facet is itself proportional to the average photon density of the mode. This is the case because, as explained in Chapter 1, the intensity of any wave (forward or backward propagating) at any location follows from the intensity of one wave at one facet by purely linear processes such as amplification and reflection. The spontaneous emission rate is independent of the photon density (but depends on the carrier density). The photon density rate equation can thus be written as [5]

$$\frac{dS_m}{dt} = [G(N, \lambda_m) - \gamma_m] S_m + \frac{R_{sp}}{V_{act}} \quad (2.3)$$

with R_{sp} the total number of spontaneously emitted photons that couple into the considered mode per unit time, S_m the average photon density (in cm^{-3}) in the mode, and γ_m the total loss of the mode m. This loss is often decomposed in internal absorption and mirror loss as follows

$$\gamma_m = \gamma_{int} + \gamma_{fac,m} = v_g(\alpha_{int} + \alpha_{fac,m}) \quad (2.4)$$

with α_{int} the internal absorption (per centimeter, typically 10 to 50 cm^{-1}) and $\alpha_{fac,m}$ the total facet loss (through both facets) of mode m. $\gamma_{fac,m} S_m V_{act}$ is the number of photons lost through both facets per time unit. It can be noticed that a part of α_{int}, the contribution from free carrier and intervalence band absorption, also depends on the carrier density.

2.3 PHASE EQUATIONS

The phase equations are not rate equations in the sense of equations describing the rate of change in time of a number (or density) of particles. They just express the phase resonance condition for the different cavity modes. They are also not as general as the previous rate equations (which are valid for all lasers, provided the averaging is done properly), but this will become clearer in Chapter 4. Here we will derive the commonly used phase equations, which only hold exactly for Fabry-Perot laser diodes.

The resonance requires that the fields have undergone a total phase shift equal to a multiple of 2π after one complete round trip inside the cavity (see, for example, the round-trip condition explained in Chapter 1). For a cavity mode m with an effective refractive index n_e (which predominantly depends on the carrier density N and on the wavelength λ_m or the frequency ω_m), this condition can be expressed as

$$2 \frac{\omega_m}{c} n_e(N, \omega_m) L = 2m\pi \quad (2.5)$$

with m an integer number and L the cavity length. The small-signal (or linearized) form of this equation reduces to

$$\Delta\omega_m n_e + \omega_m \frac{\partial n_e}{\partial \omega} \Delta\omega_m + \omega_m \frac{\partial n_e}{\partial N} \Delta N = 0 \tag{2.6}$$

or, in a more often used form,

$$\Delta\omega_m = -\frac{\omega_m}{c} v_g \frac{\partial n_e}{\partial N} \Delta N = \frac{\alpha}{2} \frac{\partial G(N, \lambda_m)}{\partial N} \Delta N \tag{2.7}$$

The quantity α introduced (and defined) in the last expression is the so-called linewidth enhancement factor [6]. It mainly depends on the wavelength and usually has a value between 1 and 5.

2.4 INTRODUCING NOISE IN THE RATE EQUATIONS

The spontaneous emission as it is included in the rate equation (2.3) for the photon densities only represents the average rate of spontaneously emitted photons that couple into the considered cavity mode. In reality, spontaneous emission is a random process occurring at random instants and resulting in photons that have a random phase with respect to the coherent laser light. The randomness can, however, be taken into account by including so-called Langevin functions in the rate equation for the photon densities and in the phase equations [7]

$$\begin{aligned}\frac{dS_m}{dt} &= [G(N, \lambda_m) - \gamma_m] S_m + \frac{R_{sp}}{V_{act}} + F_{S,m}(t) \\ \Delta\omega_m &= \frac{\alpha}{2} \frac{\partial G(N, \lambda_m)}{\partial N} \Delta N + F_{\varphi,m}(t)\end{aligned} \tag{2.8}$$

with $F_{S,m}$ and $F_{\varphi,m}$ representing stochastic processes with zero average. To derive the second-order moments of these Langevin functions, it is convenient to express the time-dependent electric field E_m of mode m in terms of the number of photons $S_m V_{act}$ and the phase φ_m [8]

$$E_m(t) = \sqrt{S_m V_{act}} \exp[j\varphi_m(t)] \tag{2.9}$$

A phasor diagram of this laser field is shown in Figure 2.2. Each spontaneous emission (at instant t_i) obviously results in one photon and hence in an extra field (with random phase θ_i and with magnitude 1) that must be added to the existing laser field.

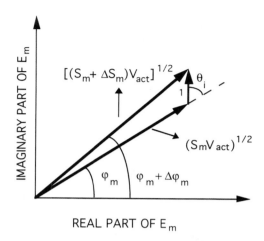

Figure 2.2 Influence of a spontaneous emission on the complex field.

From Figure 2.2, it easily follows that the resulting changes in magnitude and phase of the laser field are given by

$$\Delta(S_m V_{act}) = 1 + 2\sqrt{S_m V_{act}} \cos \theta_i$$

$$\Delta\varphi_m = \frac{1}{\sqrt{S_m V_{act}}} \sin \theta_i \tag{2.10}$$

with θ_i being uniformly distributed between 0 and 2π. The changes in photon density and in phase resulting from all spontaneous emissions can be written as

$$\Delta S_m = \frac{1}{V_{act}} \sum_i (1 + 2\sqrt{S_m V_{act}} \cos \theta_i) \, \delta(t - t_i)$$

$$\Delta\varphi_m = \frac{1}{\sqrt{S_m V_{act}}} \sum_i \sin \theta_i \, \delta(t - t_i) \tag{2.11}$$

with t_i the randomly distributed instants at which spontaneous emissions occur. There are on average R_{sp} such instants per unit of time, with the average time between consecutive spontaneous emissions being on the order of 1 ps or less. Hence, averaging of the above equations over a few tens of picoseconds (or limiting the timescale of the rate equations to this time) gives us

$$\frac{1}{\tau}\int_{t}^{t+\tau}\Delta S_m(t')dt' = \frac{1}{V_{act}}\frac{1}{\tau}\int_{t}^{t+\tau}dt'\left\{\sum_i (1 + 2\sqrt{S_m V_{act}}\cos\theta_i)\,\delta(t'-t_i)\right\}$$

$$= \frac{R_{sp}}{V_{act}} + F_{S,m}(t)$$

$$\frac{1}{\tau}\int_{t}^{t+\tau}\Delta\varphi_m(t')dt' = \frac{1}{\sqrt{S_m V_{act}}}\frac{1}{\tau}\int_{t}^{t+\tau}dt'\sum_i \sin\theta_i\,\delta(t'-t_i) = F_{\varphi,m}(t) \quad (2.12)$$

The first term on the right-hand side of (2.12) is exactly the term R_{sp}/V_{act} that is already present in (2.3). For the second-order moments of the Langevin functions, one finds

$$\langle F_{S,m}(t)\,F_{S,m}(t')\rangle = \frac{2R_{sp}S_m}{V_{act}}\delta(t-t')$$

$$\langle F_{\varphi,m}(t)\,F_{\varphi,m}(t')\rangle = \frac{R_{sp}}{2S_m V_{act}}\delta(t-t') \quad (2.13)$$

while the Langevin functions for phase and photon density equations as well as the Langevin functions corresponding with different modes are uncorrelated.

The rate equation for the carrier density must also be extended with a Langevin function F_c

$$\frac{dN}{dt} = \frac{J}{qd} - AN - BN^2 - CN^3 - \sum_m G(N,\lambda_m)S_m + F_c(t) \quad (2.14)$$

A part of this Langevin function has its origin in the spontaneous emission and is thus related to the functions $F_{S,m}$. However, spontaneous emissions induce opposite changes in photon number and in carrier number (a spontaneous emission increases the photon number by 1 and decreases the carrier number by 1). We can therefore write

$$F_c(t) = F_N(t) - \sum_m F_{S,m}(t) \quad (2.15)$$

in which F_N represents the shot noise related to the spontaneous carrier recombination. This shot noise is not correlated with the spontaneous emission noise (which basically is shot noise associated with the stimulated emission and absorption processes) and its second-order moment can be derived from the standard formula for shot noise [9]. One finds

$$\langle F_N(t) \, F_N(t') \rangle = \frac{2}{V_{act}} (AN + BN^2 + CN^3) \, \delta(t - t') \qquad (2.16)$$

2.5 OPTICAL GAIN AND ABSORPTION

The optical gain in the active layer expresses the amplification and is a measure for the strength of the net stimulated emission. It is caused by transitions of electrons from the conduction to the valence band and vice versa. Since photons possess a rather small momentum, such transitions are only possible between energy states (at energy E_2 above the conduction band edge and energy E_1 below the valence band edge) with essentially the same momentum k for the carriers.

Different approaches for the calculation of gain as a function of photon energy and carrier density have already been proposed. Here we outline a simplified approach based on the assumption that the recombining carriers in conduction and valence bands possess an identical momentum. For parabolic bands [1], this momentum can be easily related to the energies E_2 and E_1 in the conduction and valence bands, respectively:

$$k = \frac{\sqrt{2m_c E_2}}{(h/2\pi)} = \frac{\sqrt{2m_v E_1}}{(h/2\pi)} \qquad (2.17)$$

with m_c and m_v the effective masses of carriers in conduction and valence bands and h Planck's constant. As a consequence, there is a unique correspondence between the photon energy $E \, (= E_g + E_1 + E_2)$ and the energy levels in conduction and valence bands. The stimulated emission rate at the photon energy E is then proportional to:

- The light intensity at energy E;
- The density of states ρ_{red} (equal for conduction and valence bands because of (2.19)) at the energy levels under consideration;
- The probability f_c that the state at E_1 is occupied;
- The probability $1 - f_v$ that the state at E_2 is empty.

f_c and f_v are respectively, the Fermi-Dirac distribution functions for conduction and the resp. valence band. A similar dependence can be used for the absorption rate.

The net result of both stimulated emission and absorption is a gain g during propagation. With s being the coordinate along the propagation direction, the optical power obeys [10]

$$\frac{dP}{ds} = gP \qquad (2.18)$$

in which the material gain g is given by [11]

$$g(N, E) = K(E)\rho_{\text{red}}(E)[f_c(E_2) - f_v(E_1)]$$

$$K(E) = \frac{q^2 h |M|^2}{2\varepsilon_0 n c m_0^2 E}, \quad E_{1,2} = \frac{m_{c,v}}{m_c + m_v}(E - E_g) \qquad (2.19)$$

with m_0 the free electron mass, n the refractive index, ε_0 the dielectric permittivity of the vacuum, and M the transition matrix element. The dependence of g on N obviously results from the dependence of the Fermi levels on $N = P$. The reduced density of states ρ_{red} is given by [11]

$$\rho_{\text{red}}(E) = 4\pi \left(\frac{2 m_c m_v}{(m_c + m_v) h^2} \right)^{3/2} \sqrt{E - E_g} \qquad (2.20)$$

The gain G appearing in the rate equations (2.2) and (2.3) is proportional to this material gain g

$$G(N, \lambda) = \Gamma g(N, \lambda) v_g = g_{\text{mod}} v_g \qquad (2.21)$$

The confinement factor Γ accounts for the fact that only a fraction Γ of the optical power (or the photon number) is confined to the active layer and undergoes stimulated emission. $g_{\text{mod}} = \Gamma g$ is the modal gain and expresses the amplification of the intensity of the mode per unit of propagated distance. The group velocity v_g comes into the expression after the transformation to an amplification per unit of time.

It must be remarked that light propagating in the active layer also induces transitions of electrons between two valence bands (e.g., from heavy hole band to light hole or split-off band) or inside the valence and conduction bands. The intervalence band absorption [12,13] and the stimulated emission are illustrated in a typical band diagram in Figure 2.3. Obviously, intervalence band transitions are only possible for carriers with a large momentum k, and the calculation of the resulting absorption requires an accurate description of the bands at this large momentum. In particular, nonparabolicity and dependence on the k-direction have to be taken into account. The free carrier absorption [14] that is caused by transitions within the conduction or valence band is obviously not possible without violation of the conservation of momentum. The considered processes are indirect and also involve the absorption or emission of phonons. Both intervalence and free carrier absorption, however, increase with carrier density, often in a linear way.

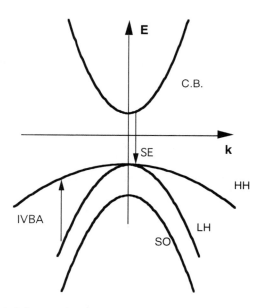

Figure 2.3 Transition of a hole between heavy hole and split-off band causing intervalence band absorption (IVBA) and transition of an electron from conduction to valence band, causing stimulated emission (SE).

The gain and free carrier or intervalence band absorption caused by the injection of carriers act as an imaginary refractive index and are thus accompanied by a change in the real part of the refractive index as well [15]. The change in the real part of the refractive index can be calculated using the Kramers-Krönig relations

$$\chi_R(E) = \frac{1}{\pi} \int_0^\infty \frac{\chi_I(E') dE'}{E - E'} \qquad (2.22)$$

This general relation between real and imaginary parts of the electric susceptibility (χ_R, and χ_I, respectively) holds if the reaction of the polarization on an electric field is passive and causal [15].

2.5.1 Bulk Materials

Both experimental and theoretical studies indicate that the gain and the refractive index in bulk material vary linearly with the carrier density

$$g(N, E) = a(E)N - b(E) \qquad (2.23)$$

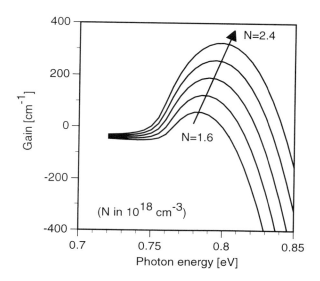

Figure 2.4 Gain in a bulk material as a function of photon energy for different levels of the electron density [16].

$a(E)$ is commonly called the *differential gain*. The gain of 1.55-μm bulk InGaAsP as a function of photon energy and carrier density is shown in Figure 2.4 [16]. The dependence on photon energy is almost quadratic near the gain peak.

Due to the temperature dependence of the Fermi-Dirac functions and of the bandgap energy in particular, the gain and refractive index depend on the temperature as well. This dependence is of little influence on the dynamic behavior of laser diodes at frequencies above 1 MHz and we will therefore not use it in the rest of this book. The doping level of the active layer is another factor that should be considered in the calculation of a and b. P-doping in general results in an increased differential gain $a(E)$, while N-doping in general results in a decreased transparency carrier density $N_t(E) = b(E)/a(E)$. a is typically 1 to 5×10^{-16} cm^2 and N_t is typically 0.5 to 2×10^{18} cm^{-3} near the gain peak energy. Figure 2.5 shows the refractive index variations corresponding to the gain of Figure 2.4. They have a linear dependence on N as well (typically -1.5×10^{-20} cm^3).

2.5.2 Quantum Wells

Since the late 1970s, active layers with very small dimensions (on the order of a few to tens of nanometers) in the transverse direction have been introduced [17–20]. These quantum wells are characterized by discrete values of the kinetic energy for motion in the (transverse) x-direction (Figure 2.6). The carriers in such structures are

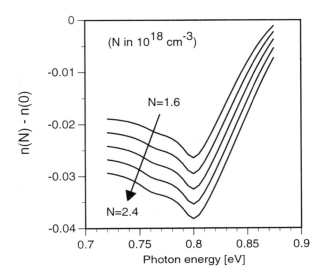

Figure 2.5 Refractive index in a bulk material as a function of photon energy for different levels of the electron density [16].

free to move in the y,z-plane and are confined in the x-direction. The reduced density of states is therefore modified to a stepwise constant function, which in turn results in a changed relation between carrier density and quasi-Fermi levels and in a modification of the expression (2.19) for the gain.

Theoretical and experimental results indicate that the gain in quantum wells is rather sublinear as a function of the carrier density. For a fixed wavelength or photon energy, the following approximation can be used:

$$g = g_0' \ln\left[\frac{N + N_s}{N_0 + N_s}\right] \tag{2.24}$$

and if only positive gain is considered,

$$g = g_0 \ln \frac{N}{N_0} \tag{2.25}$$

The gain as a function of photon energy and carrier density is shown in Figure 2.7

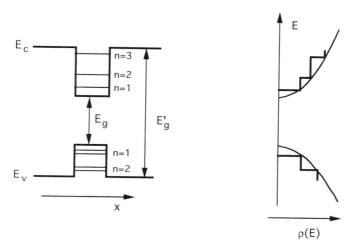

Figure 2.6 Energy levels in a quantum well and their density of states.

[21]. The sublinear function (2.25) is important in the optimization of the resonance frequency and will be considered further in Chapters 7 and 8.

2.5.3 Strained-Layer Quantum Wells

Strain can be induced in quantum wells by growing wells and barriers with a slightly different lattice constant. The strain primarily has influence on the dispersion relation for the valence bands. It causes a shift in energy between the heavy and light hole bands and modifies the inplane effective masses of these bands. Compressive strain is present when the quantum well has a larger lattice constant than the barriers. Gain results from electron transitions between the conduction band and the heavy hole band in this case. The effective mass and the density of states of the heavy hole band are reduced considerably, however. Therefore, a smaller carrier density is required to reach transparency and a higher maximum gain and differential gain are possible in such cases [23–25].

Tensile strain is present when the quantum well has a smaller lattice constant than the barriers. Lasing is now caused by electron transitions between conduction and light hole bands, but the effective mass and hence the density of states of the light hole band increases considerably with the strain. In spite of this large effective mass, it has however been found that tensile strain results in a lower threshold current, a higher differential gain, and a lower linewidth enhancement factor than compressive strain.

The intervalence band absorption and its dependence on carrier density are a

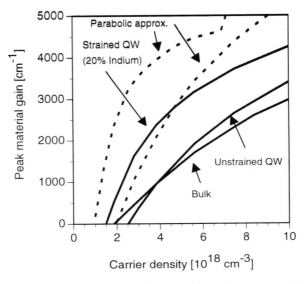

Figure 2.7 Gain as a function of photon energy and carrier density for quantum well structures [22].

lot smaller in strained quantum wells than in bulk or unstrained quantum wells. The strain-induced deformation of the heavy or light hole band causes the intervalence band transitions to occur at much larger momentum values for which the occupation probabilities are a lot smaller.

2.5.4 Gain Suppression

It has been assumed so far that the gain only depends on the carrier density and the wavelength and not on the optical power level or the photon density. This is true as long as the carrier distribution in conduction and valence bands resembles the distribution for thermal equilibrium.

In lasers biased far above threshold, however, high power levels are present at one or a few discrete wavelengths. Hence, the stimulated emission or absorption will cause the recombination or excitation of carriers at particular energy levels in the bands. This results in an initial disturbance of the thermal equilibrium distribution, as shown schematically in Figure 2.8. This deviation, which is called *spectral hole burning*, is canceled after a very short but finite time τ_{shb} (when the empty states are filled with electrons from neighboring states through intraband collisions), but still affects the gain properties of the material. The influence on the gain can be derived from the following rate equation for the fraction N_E of the carrier density that contributes to the gain at photon energy E

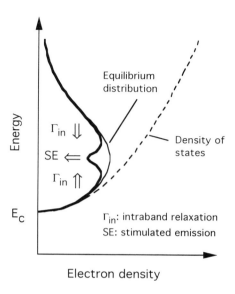

Figure 2.8 Disturbance of the equilibrium distribution for the electrons in the conduction band due to spectral hole burning. (*Source:* [17], © IEEE 1984.)

$$\frac{dN_E}{dt} = -\frac{N_E - N_{E0}}{\tau_{shb}} - \Gamma v_g g(E) S \qquad (2.26)$$

with S the photon density at energy E and N_{E0} the value of N_E for $S = 0$. Using a linear dependence of g on N_E (at least in the vicinity of N_{E0}), one finds for the gain

$$\frac{\partial g}{\partial t} = -\frac{g - g(S = 0)}{\tau_{shb}} - \Gamma \frac{\partial g}{\partial N} v_g g S \qquad (2.27)$$

and on a timescale that is large compared with τ_{shb},

$$g(N, E, S) = \frac{g(N, E, 0)}{1 + \xi_{shb} S} \text{ with } \xi_{shb} = \Gamma v_g \frac{\partial g}{\partial N} \tau_{shb} \qquad (2.28)$$

Typical values of τ_{shb} are in the range 10^{-14} to 10^{-13} sec [26].

A similar gain suppression is caused by carrier heating. This process has its origin in the stimulated recombination of carriers with an energy below the quasi-Fermi level. The loss of such "cold" carriers results in a nonthermal distribution that is restored by an increase in the temperature T_c of the remaining carriers, described by the following equation

$$\frac{dT_c}{dt} = -\frac{T_c - T_1}{\tau_h} + \frac{F_c - F_v - h\nu}{C_h}\Gamma v_g g S \qquad (2.29)$$

with T_1 the lattice temperature, τ_h the restoration time, and C_h the heat capacity of the carriers. F_c and F_v are the quasi-Fermi levels of conduction and valence bands. Assuming a linear dependence of the gain on T_c and on a timescale that is large compared with τ_h, one finds the following gain suppression

$$g(N, E, S) = \frac{g(N, E, 0)}{1 + \xi_h S} \text{ with } \xi_h = \frac{F_c - F_v - h\nu}{C_h}\Gamma v_g \frac{\partial g}{\partial T_c}\tau_h \qquad (2.30)$$

Typical values of τ_h are in the range 10^{-13} to 10^{-12} sec [27].

The exact contribution of carrier heating and spectral hole burning is not very well known [28,29]. Their combined effect can, however, still be expressed as

$$g(N, E, S) = \frac{g(N, E, 0)}{1 + \xi S} \qquad (2.31)$$

or linearized

$$g(N, E, S) = g(N, E, 0)(1 - \xi S) \qquad (2.32)$$

In the case of multimode lasers, with modes at energies E_k and with photon densities S_k, this expression can be generalized to

$$g(N, E_m, S_k) = g(N, E_m)\left\{1 - \sum_k \xi_{mk} S_k\right\} \qquad (2.33)$$

with the coefficients ξ_{mk} depending on both the energies E_m and E_k. Typical values for ξ are 10^{-18} to 10^{-17} cm^3.

2.6 SOME WELL-KNOWN SOLUTIONS OF THE RATE EQUATIONS

While a numerical solution of the rate equations can be obtained under all circumstances, it is also relatively easy to derive simple expressions for several important laser characteristics from analytical, approximative solutions. Here we will give an overview of these well-known expressions and their derivations. A more detailed account of analytical expressions and their significance for design purposes will be given in the chapters dealing with specific characteristics. In all the derivations, it is assumed that the cavity loss is independent of the optical power level and that the spontaneous emission rate is independent of the wavelength and equal to the spontaneous emission rate at the lasing wavelength.

Before proceeding with the analytical solutions, we summarize the different rate equations and the meaning of the different quantities appearing in them one more time in Table 2.1.

Table 2.1
Overview of the Different Rate Equations

Photon Density Equations

$$\frac{dS_m}{dt} = [G(N, \lambda_m) - \gamma_m] S_m + \frac{R_{sp}}{V_{act}} + F_{S,m}(t)$$

with

$$\langle F_{S,m}(t) F_{S,n}(t') \rangle = \frac{2 R_{sp} S_m}{V_{act}} \delta(t - t') \delta_{mn}$$

m,n = mode number
V_{act} = volume of active layer
S_m = average photon density of mode m
R_{sp} = spontaneous emission rate
γ_m = internal and mirror loss of mode m
$G(N, \lambda_m)$ = gain (per unit time) of mode m
$F_{S,m}$ = Gaussian stochastic process with zero average

Carrier Density Equations

$$\frac{dN}{dt} = \frac{J}{qd} - AN - BN^2 - CN^3 - \sum_m G(N, \lambda_m) S_m$$

$$+ F_N(t) - \sum_m F_{S,m}(t)$$

with

$$\langle F_N(t) F_N(t') \rangle = \frac{2}{V_{act}} (AN + BN^2 + CN^3) \delta(t - t')$$

N = average carrier density
J = injected current density
d = thickness of the active layer
A = monomolecular recombination coefficient
B = bimolecular recombination coefficient
C = Auger recombination coefficient
F_N = Gaussian stochastic process with zero average

Phase Equations

$$\Delta \omega_m = \frac{\alpha}{2} \frac{\partial G(N, \lambda_m)}{\partial N} \Delta N + F_{\varphi,m}(t)$$

with

$$\langle F_{\varphi,m}(t) F_{\varphi,n}(t') \rangle = \frac{R_{sp}}{2 S_m V_{act}} \delta(t - t') \delta_{mn}$$

$\Delta \omega_m$ = variation of the optical pulsation of mode m
α = linewidth enhancement factor
ΔN = variation of the average carrier density
$F_{\varphi,m}$ = Gaussian stochastic process with zero average

2.6.1 The Static Side-Mode Suppression

In a laser biased sufficiently far above the threshold current, one can assume that the gain of the lasing mode is clamped to a value that equals the total loss of the cavity ($G = \gamma_m$). The gain clamping follows readily from (2.3) if static operation is assumed and if $S_m \gg R_{sp}/V_{act}$. From the same equation applied on a side mode s, one finds for the average photon density of the side mode

$$S_s = \frac{R_{sp}}{V_{act}[\gamma_s - G(N, \lambda_s)]} \qquad (2.34)$$

with γ_s the total loss and $G(N, \lambda_s)$ the gain of the side mode at wavelength λ_s. The wavelength dependence of the gain is always much weaker than that of the loss γ in DFB lasers, and therefore $G(N, \lambda_s)$ can be approximated by $G(N, \lambda_m) \approx \gamma_m$. The side-mode suppression can be approximated as S_m/S_s (assuming an identical ratio between output power and average photon density for both main and side mode) and expressed as

$$\text{SMSR} = \frac{V_{act}[\gamma_s - G(N, \lambda_s)]}{R_{sp}} S_m \approx \Delta(\Gamma g_{th})L \frac{P}{\gamma_m n_{sp} h\nu} \qquad (2.35)$$

with P the average power density in the cavity and $\Delta(\Gamma g_{th})$ the difference in threshold gain between main and side modes. The second expression of (2.35) follows from the relations $\gamma_s - G = \gamma_s - \gamma_m = v_g \Delta(\Gamma g_{th})$ and $P = h\nu wd v_g S_m$. It can be noticed that the side mode reaches threshold when the gain equals γ_s. The difference $\gamma_s - G = \gamma_s - \gamma_m$ is therefore also the difference in threshold gain G_{th} between mode m and mode s.

By using typical values in (2.35), such as $n_{sp} = 2$, $\gamma_m = 100$ cm^{-1}, and an average intracavity power of 2 mW (which, for cleaved facets, corresponds to 1-mW output power), we find that a static side-mode suppression of 30 dB requires a normalized threshold gain difference $\Delta(\Gamma g_{th})L$ of ±0.1. As will be discussed in Chapter 3, the threshold gain difference in DFB lasers depends in general on the specific longitudinal structure and even on the bias level. Its value, calculated at threshold, can then only give an indication of the single-mode behavior.

2.6.2 FM and AM Behavior

The AM and FM response can be obtained after linearization of (2.2) and (2.3). We restrict ourselves here to truly single-mode lasers with a side-mode suppression of 30 dB or more and ignore the mode index m from now on. Side modes have no impact on the ac response of the main mode in this case [30]. We also neglect the wavelength dependence of the gain. By introducing the expansions

$$N(t) = N_0 + \text{Re}(N_1 e^{j\Omega t})$$
$$S(t) = S_0 + \text{Re}(S_1 e^{j\Omega t})$$
$$J(t) = J_0 + \text{Re}(J_1 e^{j\Omega t}) \tag{2.36}$$

and linearizing (2.2) and (2.3) in N_1 and S_1, we find the small-signal equations

$$\left(j\Omega + \frac{1}{\tau_d} + \frac{\partial G}{\partial N} S_0 \right) N_1 = \frac{J_1}{qd} - \left(G + \frac{\partial G}{\partial S} S_0 \right) S_1$$

$$\left(j\Omega - \frac{\partial G}{\partial S} S_0 + \frac{R_{sp}}{V_{act} S_0} \right) S_1 = \left(\frac{\partial G}{\partial N} - \frac{\partial \gamma}{\partial N} \right) S_0 N_1 \tag{2.37}$$

in which τ_d denotes the differential or small-signal carrier lifetime

$$\left(\frac{1}{\tau_d} \right) = A + 2B N_0 + 3 C N_0^2 \tag{2.38}$$

These equations can easily be solved for S_1 (which gives the AM response) and N_1 (which through (2.7) gives the FM response). They can also be easily extended to the multimode case. One finds the formulas

$$S_1 = \frac{\left(\frac{\partial G}{\partial N} - \frac{\partial \gamma}{\partial N} \right) S_0 \frac{J_1}{qd}}{\left\{ \left(j\Omega - \frac{\partial G}{\partial S} S_0 + \frac{R_{sp}}{V_{act} S_0} \right) \left(j\Omega + \frac{1}{\tau_d} + \frac{\partial G}{\partial N} S_0 \right) + G \left(\frac{\partial G}{\partial N} - \frac{\partial \gamma}{\partial N} \right) S_0 \right\}}$$

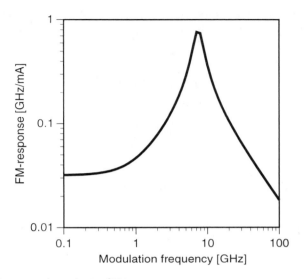

Figure 2.9 Typical frequency dependence of FM response.

$$\Delta\omega = \frac{\dfrac{\alpha}{2}\dfrac{\partial G}{\partial N}\left(j\Omega - \dfrac{\partial G}{\partial S}S_0 + \dfrac{R_{sp}}{V_{act}S_0}\right)\dfrac{J_1}{qd}}{\left\{\left(j\Omega - \dfrac{\partial G}{\partial S}S_0 + \dfrac{R_{sp}}{V_{act}S_0}\right)\left(j\Omega + \dfrac{1}{\tau_d} + \dfrac{\partial G}{\partial N}S_0\right) + G\left(\dfrac{\partial G}{\partial N} - \dfrac{\partial \gamma}{\partial N}\right)S_0\right\}}$$

(2.39)

It must be noticed that the actual AM response further depends on the relationship between output power and photon number S. Equation (2.39), however, gives the exact modulation frequency dependence of the AM response. The gain G is equal to the threshold gain $G_{th} = G(N_{th})$. The typical frequency dependence of AM and FM response is shown in Figures 2.9 and 2.10.

The modulation frequency dependence in the denominator of (2.39) gives rise to a resonance phenomenon, better known as the relaxation oscillation. This resonance is also subject to damping, mainly caused by spontaneous emission and gain suppression. The resonance frequency f_r and damping θ of the relaxation oscillation are given by [31]

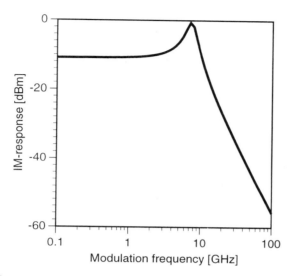

Figure 2.10 Typical frequency dependence of AM response.

$$(2\pi f_r)^2 = G_{th}\left(\frac{\partial G}{\partial N} - \frac{\partial \gamma}{\partial N}\right)S_0$$

$$\theta = \frac{1}{\tau_d} + \frac{R_{sp}}{V_{act}S_0} + \frac{\partial G}{\partial N}S_0 - \frac{\partial G}{\partial S}S_0 \quad (2.40)$$

The contribution of the spontaneous emission to the damping is only relevant at small (subthreshold) power levels and is usually neglected above threshold. Without this term, the resonance frequency and damping are seen to increase with increasing bias level S_0 and with increasing net differential gain. A considerable contribution to damping is given by the gain suppression $\partial G/\partial S$ (or ξ). Typical values of the resonance frequency are on the order of magnitude of a few gigahertz.

The relaxation oscillations describe the resonant exchange of energy between photons and carriers (electrons) via the processes of stimulated emission and absorption. The oscillation can be described by assuming that a surplus of electrons with respect to the equilibrium exists. The gain exceeds the loss in this case and the stimulated emission creates more and more photons, resulting in an increasing stimulated emission rate. This in turn depletes the carriers inside the cavity, so that eventually the gain can no longer overcome the loss. At this point, the number of photons decreases again, the stimulated emission rate decreases, and the carrier density increases again above its equilibrium value, where a new cycle starts. This picture also shows why spontaneous emission and carrier recombination (which prevent carriers and photons from contributing to the stimulated recombination) or gain suppression (which restricts the increase of stimulated recombination for increasing photon numbers) cause a damping of the oscillation. It also shows why the resonance frequency increases with increasing net differential gain and with increasing photon density S_0. The larger the photon density or the differential gain, the faster the stimulated emission rate changes with carrier density according to the second equation of (2.37).

From (2.39), it also follows that the static (i.e., at modulation frequencies of 1 MHz) AM response, which is an indication of the internal efficiency, is proportional to the inverse threshold gain. The static FM response, on the other hand, is, at practical power levels, mainly proportional to the linewidth enhancement factor and the gain suppression coefficient.

2.6.3 Harmonic Distortion Characteristics

Harmonic distortion in the output power of modulated lasers is caused by different nonlinear effects such as gain suppression or leakage currents. Here we will illustrate the influence of the intrinsic nonlinearity caused by the coupling of carrier density and photon density equations. The distortion can be calculated from a higher order small-signal analysis using (2.2) and (2.3) and the following expansion [32]

$$J(t) = J_0 + \mathrm{Re}(J_1 e^{j\Omega t})$$

$$N(t) = N_0 + \mathrm{Re}\{N_1 e^{j\Omega t} + N_2 e^{2j\Omega t} + N_3 e^{3j\Omega t}\}$$

$$S(t) = S_0 + \mathrm{Re}\{S_1 e^{j\Omega t} + S_2 e^{2j\Omega t} + S_3 e^{3j\Omega t}\} \tag{2.41}$$

The analysis proceeds as follows. First the first-order quantities are calculated using a linearization of the rate equations. The expansions (2.41) up to second order are then again substituted in the exact rate equations, in which now the dependence on N_1 and S_1 is approximated by a quadratic form and the dependence on N_2 and S_2 is linearized. Since the first-order quantities are known, a linear system in the second-order quantities results and its solution allows the determination of the second-order harmonic distortion. A similar procedure can be followed for the calculation of the third-order distortion. From such an analysis, one finds for the second- and third-order harmonic distortion as a function of the modulation depth $m = S_1/S_0$ [33]:

$$\frac{S_2}{S_1} = \frac{j\Omega\left(2j\Omega + \dfrac{1}{\tau_d}\right)}{\left\{2j\Omega\left(2j\Omega + \dfrac{1}{\tau_d} + \dfrac{\partial G}{\partial N}S_0\right) + \left(\dfrac{\partial G}{\partial N} - \dfrac{\partial \gamma}{\partial N}\right)G_{th}S_0\right\}} \cdot \frac{m}{2}$$

$$\frac{S_3}{S_1} = \frac{1}{2}\frac{\left(3j\Omega + \dfrac{1}{\tau_d}\right)\left(3j\Omega\dfrac{S_2}{S_1}m - \dfrac{j\Omega}{2}m^2\right)}{\left\{3j\Omega\left(3j\Omega + \dfrac{1}{\tau_d} + \dfrac{\partial G}{\partial N}S_0\right) + \left(\dfrac{\partial G}{\partial N} - \dfrac{\partial \gamma}{\partial N}\right)G_{th}S_0\right\}} \quad (2.42)$$

Gain suppression and spontaneous emission have been neglected in deriving (2.42). The second- and third-order harmonic distortion caused by the intrinsic nonlinearities are mainly important at larger modulation frequencies (typically above 500 MHz). The typical variation with modulation frequency is shown in Figure 2.11 and Figure 2.12.

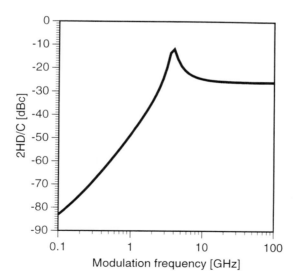

Figure 2.11 Second-order harmonic distortion caused by the intrinsic nonlinearity for constant optical modulation depth.

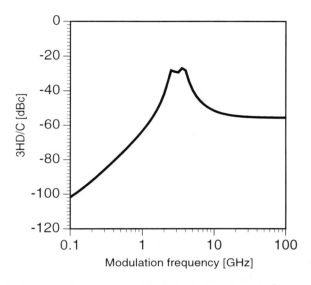

Figure 2.12 Third-order harmonic distortion caused by the intrinsic nonlinearity for constant optical modulation depth.

2.6.4 Large-Signal Characteristics

The main large-signal characteristics are related to the (digital) on/off switching of the laser diode. The typical behavior of the carrier density and the photon density under such switching is shown in Figure 2.13 and 2.14. The delay time τ_D is the time needed for the carrier density to reach the threshold value. Its value is different from zero if the laser is biased at a current I_0 below the threshold current I_{th} when at $t = 0$, the current is switched to a value I_b above the threshold current. An expression for it follows from the solution of the carrier rate equation (2.2), in which the stimulated emission is neglected. For $t > 0$, this equation reads

$$\frac{dN}{dt} = \frac{I_b}{qV_{act}} - AN - BN^2 - CN^3 \qquad (2.43)$$

and can be transformed into [33]

$$\frac{dI}{I_b - I(N)} = \frac{dI}{dN}\frac{dt}{qV_{act}} \quad \text{with } I(N) = qV_{act}(AN + BN^2 + CN^3) \qquad (2.44)$$

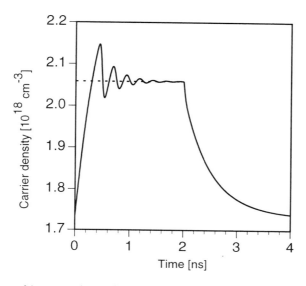

Figure 2.13 Variation of the carrier density during switch-on and switch-off (the current is 2 ns above threshold and 2 ns below threshold at the original value).

A value for the delay time then easily follows if (dI/dN) is approximated as a constant

$$\tau_D = \frac{1}{A + 2BN + 3CN^2} \ln\left(\frac{I_b - I(N(0))}{I_b - I(N_{\text{th}})}\right) = \tau_d \ln\left(\frac{I_b - I_0}{I_b - I_{\text{th}}}\right) \qquad (2.45)$$

with $I(N(0)) = I(0)$ and $I(N_{\text{th}}) = I_{\text{th}}$. τ_d is the differential carrier lifetime, which can be calculated using, for example, the carrier density at I_0 (to obtain an upper bound) or at I_{th} (to obtain a lower bound). More detailed formulas, taking into account the nonlinear dependence of the carrier lifetime on carrier density, can, however, also be derived [34].

For the calculation of the switch-on time of a laser that is originally biased slightly above threshold (again with a bias current I_{off} and with an average photon density S_{off}) and that, at $t = 0$, is switched to a current I_{on} (with an average photon density S_{on}), one still neglects the influence of stimulated emission in the carrier rate equation. (Note that S_{on} is obviously meant to be sufficiently large in digital on-/off-switching.) This is justified from the typical behavior of the photon density as a

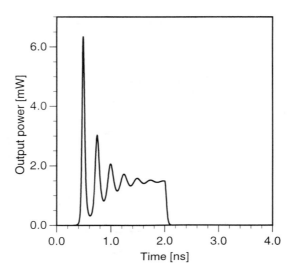

Figure 2.14 Variation of the output power during switch-on and switch-off. (The current is 2 ns above threshold and 2 ns below threshold at the original value.)

function of time during the switch-on time t_{on} (Figure 2.14). The photon density remains close to the small value S_{off} for most of the time t_{on}. The solution of the carrier rate equation then reduces to

$$N = N_{th} + \frac{(I_{on} - I_{off})t}{qV_{act}} \qquad (2.46)$$

with the carrier density at I_{off} being approximated as the threshold carrier density. With this expression for the carrier density, the photon density as a function of time can be calculated easily if the spontaneous emission and the gain suppression are neglected. By approximating the gain in the vicinity of N_{th} as a linear function of the carrier density, one finds the photon density equation

$$\frac{dS}{dt} = \left(\frac{\partial G}{\partial N} - \frac{\partial \gamma}{\partial N}\right) \frac{I_{on} - I_{off}}{qV_{act}} St = G(N_{th}) \left(\frac{\partial G}{\partial N} - \frac{\partial \gamma}{\partial N}\right)(S_{on} - S_{off})St \quad (2.47)$$

with $I_{on} - I_{off} = G_{th}(S_{on} - S_{off})V_{act}$. Equation (2.47) has the following solution

$$S(t) = S_{off} \exp\left\{\frac{1}{2} G_{th} \left(\frac{\partial G}{\partial N} - \frac{\partial \gamma}{\partial N}\right)(S_{on} - S_{off})t^2\right\} \qquad (2.48)$$

from which the turn-on time (i.e., the time when S reaches S_{on} for the first time) can be calculated

$$t_{on} = \frac{\sqrt{2}}{\omega_{on}} \left[\ln\left(\frac{S_{on}}{S_{off}}\right) \right]^{1/2} \text{ with } \omega_{on}^2 = \left[G_{th} \left(\frac{\partial G}{\partial N} - \frac{\partial \gamma}{\partial N} \right) S_{on} \right] \quad (2.49)$$

ω_{on} is the resonance pulsation corresponding to the on-state.

2.6.5 Power Spectrum, Linewidth, and Intensity Noise

The spectra of the FM noise and of the intensity noise are derived in a similar way as the ac response. The Langevin functions now act as a driving source instead of a sinusoidal current. It turns out that, at practical bias levels, the 3-dB width of the power spectrum (the linewidth) only depends on the low-frequency value of the FM noise, and therefore one can neglect the dynamics in solving the rate equations. We will also ignore the steady-state spontaneous emission here (we assume a relatively high bias level) and we consider a single-mode laser. The static fluctuations in frequency and intensity are then easily derived from (2.8). Neglecting the wavelength dependence of the gain and the loss gives

$$N_1 = -\frac{F_S}{\left(\frac{\partial G}{\partial N} - \frac{\partial \gamma}{\partial N} \right) S_0}$$

and

$$\Delta\omega = -\frac{\alpha}{2} \frac{\frac{\partial G}{\partial N}}{\left(\frac{\partial G}{\partial N} - \frac{\partial \gamma}{N} \right)} \frac{F_S}{S_0} + F_\varphi$$

$$S_1 = \frac{\left(\frac{1}{\tau_d} + \frac{\partial \gamma}{\partial N} \right) F_S + \left(\frac{\partial G}{\partial N} - \frac{\partial \gamma}{\partial N} \right) S_0 F_N}{G \left(\frac{\partial G}{\partial N} - \frac{\partial \gamma}{\partial N} \right) S_0} \quad (2.50)$$

The carrier density dependence of the loss γ is often assumed small as compared to the carrier dependence of the gain G or included in the linewidth enhancement factor α. Moreover, substitution of typical values for τ_d (10^{-9} sec) in (2.50) shows that F_N can be neglected for not too high power levels. The expressions in (2.50) are valid for frequencies up to ± 1 GHz and on a timescale of a few nanoseconds. One then finds (when leaving the mode index m behind)

$$\langle \Delta\omega(t)\Delta\omega(t')\rangle = \frac{R_{sp}}{2S_0 V_{act}}(1+\alpha^2)\delta(t-t')$$

$$\langle \Delta S(t)\Delta S(t')\rangle = \langle S_1(t')\, S_1(t')\rangle = \frac{2R_{sp} V_{act}}{\tau_d^2 G^2 \left(\frac{\partial G}{\partial N}\right)^2 S_0}\delta(t-t') \qquad (2.51)$$

The spectral density of the FM noise and of the RIN can now be calculated in a straightforward manner [35]

$$S_{\Delta\omega}(f) = \int_{-\infty}^{+\infty} dt \langle \Delta\omega(t)\Delta\omega(0)\rangle \exp(j2\pi ft) = \frac{R_{sp}}{2S_0 V_{act}}(1+\alpha^2)$$

$$\mathrm{RIN} = \frac{S_{\Delta S}(f)}{S_0^2} = \frac{2R_{sp} V_{act}}{\tau_d^2 G^2 \left(\frac{\partial G}{\partial N}\right)^2 S_0^2} \qquad (2.52)$$

The values predicted by (2.52) will, in reality, only be found at frequencies that are sufficiently below 1 GHz. In some cases, one is also interested in the spectra at higher frequencies, and a more detailed calculation, taking into account the dynamics, is then required.

The power spectrum, however, can often be calculated in an accurate way by assuming a white spectrum for the FM noise, with a spectral density given by (2.52). The RIN is also neglected in this case, and one expresses the single-mode field, emitted through one of the facets [35], as

$$E(t) = E_0 \exp\left\{-j\left[\omega_0 t + \int_0^t \Delta\omega(t')\, dt'\right]\right\} \qquad (2.53)$$

with E_0 and ω_0 being constants. The power spectrum, defined as the spectral density of the field, is given by

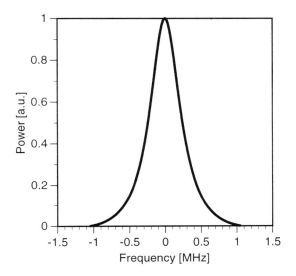

Figure 2.15 Lorentzian shape of the power spectrum.

$$S_E(f) = \int_{-\infty}^{+\infty} dt \langle E^*(t)E(0)\rangle e^{j2\pi ft}$$

$$= |E_0|^2 \int_{-\infty}^{+\infty} dt \left\langle \exp\left(j\int_0^t \Delta\omega(t')\, dt'\right)\right\rangle e^{j2\pi(f-f_0)t} \quad (2.54)$$

At this point, it must be noticed that, due to the Gaussian character of the Langevin forces, $\Delta\omega$ and its integral can also be treated as Gaussian variables (for which $\langle e^{jx}\rangle = \exp(-0.5\,\langle x^2\rangle)$). In addition, it follows that

$$\left\langle \left(\int_0^t \Delta\omega(t')\, dt'\right)^2\right\rangle = \int_0^t dt_1 \int_0^t dt_2 \langle \Delta\omega(t_1)\Delta\omega(t_2)\rangle = \frac{R_{sp}}{2S_0 V_{act}}(1+\alpha^2)|t| \quad (2.55)$$

Equation (2.54) then reduces to

$$S_E(f) = \frac{|E_0|^2}{2\pi}\, \frac{\Delta\nu}{(f-f_0)^2 + \left(\dfrac{\Delta\nu}{2}\right)^2} \quad \text{with}\quad \Delta\nu = \frac{R_{sp}}{4\pi S_0 V_{act}}(1+\alpha^2) \quad (2.56)$$

The power spectrum has a Lorentzian line form with a 3-dB width of Δv. The expression for Δv given in (2.56) is better known as Henry's formula. It predicts that the linewidth Δv decreases proportionally with the inverse power level. The spectrum (2.56) is illustrated in Figure 2.15.

2.7 THE INFLUENCE OF EXTERNAL REFLECTIONS

External reflections (e.g., originating from an optical fiber or from lenses) can change the behavior of lasers significantly and, for example, bring along an enormous increase or decrease of the linewidth and the RIN. The influence of such external reflections is very similar for Fabry-Perot and DFB lasers and will therefore not be discussed in this book. The reflections and their influence can also be eliminated by inserting an optical isolator after the collimating lens and/or by placing the coupling elements under a small angle with respect to the laser facet. We refer to [5] for more details.

References

[1] Sze, S. M., *Physics of Semiconductor Devices*, 2nd edition, New York: Wiley, 1981.
[2] Schockley, W., and W. T. Read, "Statistics of the Recombination of Holes and Electrons," *Phys. Rev.*, Vol. 87, 1952, p. 835.
[3] Olshansky, R., C. B. Su, J. Manning, and W. Powaznik, "Measurement of Radiative and Nonradiative Recombination Rates in InGaAsP and AlGaAs Light Sources," *IEEE J. Quant. Electron.*, Vol. 20, 1984, pp. 838–854.
[4] Agrawal, G. P., and N. K. Dutta, *Long-Wavelength Semiconductor Lasers*, New York: Van Nostrand Reinhold, 1986.
[5] Petermann, K., *Laser Diode Modulation and Noise*, Dordrecht: Kluwer Academic Publishers, 1988.
[6] Osinski, M., and J. Buus, "Linewidth Broadening Factor in Semiconductor Lasers: An Overview," *IEEE J. Quant. Electron.*, Vol. 23, 1987, pp. 9–24.
[7] Vahala, K., and A. Yariv, "Semiclassical Theory of Noise in Semiconductor Lasers—Part II," *IEEE J. Quant. Electron.*, Vol. 19, 1983, pp. 1101–1109.
[8] Henry, C. H., "Phase Noise in Semiconductor Lasers," *IEEE J. Lightw. Tech.*, Vol. 4, 1986, pp. 298–311.
[9] Yariv, A., *Introduction to Optical Electronics*, 3rd edition, New York: Holt, Reinhart and Winston, 1971.
[10] Casey, H., and M. B. Panish, *"Heterostructure Lasers: Part A, Fundamental Principles,"* New York: Academic Press, 1978.
[11] Thompson, G. H. B., *Physics of Semiconductor Laser Devices*, Pitman Press, 1985.
[12] Asada, M., A. R. Adams, K. E. Stubkjaer, Y. Suematsu, Y. Itaya, and S. Arai, "The Temperature Dependence of the Threshold Current of GaInAsP/InP DH Lasers," *IEEE J. Quant. Electron.*, Vol. 17, 1981, pp. 611–619.
[13] Henry, C. H., R. A. Logan, F. R. Merritt, and J. P. Luongo, "The Effect of Intervalence Band Absorption on the Thermal Behavior of InGaAsP Lasers," *IEEE J. Quant. Electron.*, Vol. 19, 1983, pp. 947–952.
[14] Seeger, K., *Semiconductor Physics*, Berlin: Springer-Verlag, 1985.

[15] Yariv, A., *Quantum Electronics*, 3rd edition, New York: Wiley, 1989.
[16] Botteldooren, D., Ph.D. thesis (in Dutch), University of Gent, Belgium, 1990.
[17] Asada, M., and Y. Suematsu, "Density-Matrix Theory of Semiconductor Lasers With Relaxation Broadening Model—Gain and Gain Suppression in Semiconductor Lasers," *IEEE J. Quant. Electron.*, Vol. 21, 1984, pp. 434–442.
[18] Van Der Ziel, J. P., R. Dingle, R. C. Miller, W. Wiegmann, and W. A. Nordland, Jr., "Laser Oscillations From Quantum States in Very Thin GaAs-$Al_{0.2}Ga_{0.8}As$ Multilayer Structures," *Appl. Phys. Lett.*, Vol. 26, 1975, pp. 463–465.
[19] Asada, M., A. Kameyama, and Y. Suematsu, "Gain and Intervalence Band Absorption in Quantum-Well Lasers," *IEEE J. Quant. Electron.*, Vol. 20, 1984, pp. 745–750.
[20] Holonyak, N., Jr., R. M. Kolbas, R. D. Dupuis, and P. D. Dapkus, "Quantum-Well Heterostructure Lasers," *IEEE J. Quant. Electron.*, Vol. 16, 1980, pp. 170–186.
[21] Zory, P. S., ed., *Quantum Well Lasers*, San Diego: Academic Press, 1992.
[22] Corzine, S. W., R. H. Yan, and L. A. Coldren, "Theoretical Gain in Strained InGaAs/AlGaAs Quantum Wells Including Valence Band Mixing Effects," *Appl. Phys. Lett.*, Vol. 57, 1990, pp. 2835–2837.
[23] Tsang, W. T., C. Weisbuch, R. C. Miller, and R. Dingle, "Current Injection GaAs-$Al_xGa_{1-x}As$ Multi-Quantum-Well Heterostructure Lasers Prepared by Molecular Beam Epitaxy," *Appl. Phys. Lett.*, Vol. 35, 1979, pp. 635–675.
[24] Thijs, P. J. A., L. F. Tiemeijer, J. J. M. Binsma, and T. Van Dongen, "Progress in Long-Wavelength Strained-Layer InGaAs(P) Quantum-Well Semiconductor Lasers and Amplifiers," *IEEE J. Quant. Electron.*, Vol. 30, 1994, pp. 477–499.
[25] Miller, B. I., U. Koren, M. G. Young, and M. D. Chien, "Strain-Compensated Strain-Layer Superlattices for 1.5µm Wavelength Lasers," *Appl. Phys. Lett.*, Vol. 58, 1991, pp. 1952–1954.
[26] Joindot, I., and J.-L. Beylat, "Intervalence Band Absorption Coefficient Measurements in Bulk Layer, Strained and Unstrained Multiquantum Well 1.55µm Semiconductor Lasers," *Electron. Lett.*, Vol. 29, 1993, pp. 604–606.
[27] Hall, K. L., P. Y. Lai, E. P. Ippen, G. Eisenstein, and U. Koren, "Femtosecond Gain Dynamics and Saturation Behaviour in InGaAsP Multiple Quantum Well Optical Amplifiers," *Appl. Phys. Lett.*, Vol. 57, 1990, pp. 2888–2890.
[28] Kesler, M. P., and E. P. Ippen, "Subpicosecond Gain Dynamics in GaAlAs Laser Diodes," *Appl. Phys. Lett.*, Vol. 51, 1987, pp. 1765–1767.
[29] Tiemeijer, L. F., "Effects of Nonlinear Gain on Four-Wave Mixing and Asymmetric Gain Saturation in a Semiconductor Laser Amplifier," *Appl. Phys. Lett.*, Vol. 59, 1991, pp. 499–501.
[30] Tucker, R,. "High-Speed Modulation of Semiconductor Lasers," *IEEE J. Lightw. Tech.*, Vol. 3, 1985, pp. 1180–1192.
[31] Olshansky, R., P. Hill, V. Lanzisera, and W. Powazinik, "Frequency Response of 1.3µm InGaAsP High Speed Semiconductor Lasers," *IEEE J. Quant. Electron.*, Vol. 23, 1987, pp. 1410–1418.
[32] Lau, K., and A. Yariv, "Intermodulation Distortion in a Directly Modulated Semiconductor Injection Laser," *Appl. Phys. Lett.*, Vol. 45, 1984, pp. 1034–1036.
[33] Morthier, G., K. David, P. Vankwikelberge, and R. Baets, "Theoretical Investigation of the 2nd Order Harmonic Distortion in the AM Response of 1.55µm F-P and DFB Lasers," *IEEE J. Quant. Electron.*, Vol. 27, 1991, pp. 1990–2002.
[34] Konnerth, K., and C. Lanza, "Delay Between Current Pulse and Light Emission of a GaAs Injection Laser," *Appl. Phys. Lett.*, Vol. 4, 1964, pp. 120–123.
[35] Henry, C. H., "Theory of the Phase Noise and Power Spectrum of a Single Mode Injection Laser," *IEEE J. Quant. Electron.*, Vol. 19, 1983, pp. 1391–1397.

Chapter 3

Coupled-Mode Theory of DFB Laser Diodes

In the previous chapter a first type of laser diode model based on lumped rate equations was presented. In this chapter a more accurate model based on the coupled-mode theory is described.

A laser diode is a complex system in which electrical, electro-optic, optical, and thermal phenomena constantly interact in a nonlinear way. Describing all these phenomena in full detail may lead to a theoretical model that is far too complex to work with. On the other hand, introducing too much simplification may jeopardize the utility of a model. Therefore, the right trade-off is required. This depends on the complexity of the laser structure and on the laser characteristics being investigated. For DFB lasers, the coupled-mode theory just strikes this balance between simplicity and accuracy.

First of all, our theoretical model takes into account the longitudinal (i.e., axial) variations of the optical intensity and the carrier density. In DFB lasers this is important, especially to demonstrate the impact of spatial hole burning on the device behavior. The term spatial hole burning refers to the variation of the carrier density along the laser cavity.

Secondly, our longitudinal coupled-mode theory also takes the dynamic and noise behavior of the laser diode into account. Moreover, it can adequately model multisection lasers and lasers with chirped gratings. In this way, the static (continuous wave), noise, and dynamic behavior of complex DFB laser diode structures can be analyzed in detail. The theory explained here extends considerably beyond the traditional textbook material on the coupled-mode theory of DFB lasers.

This extended coupled-mode theory emphasizes the carrier dynamics in the active layer and the optical propagation in the waveguide of the laser. To this end, carrier equations and Maxwell's theory are used to establish a dynamic (t) and longitudinal (z) model. The carrier injection process in the active layer is not included in the coupled-mode model. It is discussed in Chapter 5 because this process can easily be decoupled from the carrier and photon dynamics in the active layer by assuming pure current sources as drivers for the laser.

3.1 THE PHYSICAL PROCESSES INSIDE A LASER DIODE

In Chapter 2, laser diodes are described by means of lumped rate equations. Deriving them requires several approximations. To help the reader understand these approximations, this chapter will present a thorough model for DFB lasers from which the rate equations can then be derived (see Chapter 4). However, this detailed longitudinal model, based on the coupled-mode theory, is not derived only for the sake of explaining the rate equations. The longitudinal model allows the clarification of many longitudinal spatial hole burning effects that cannot easily be covered by the rate equations.

Before deriving the coupled-mode model, it is useful to discuss the different physical processes that occur in a laser diode in some wider perspective than the view offered by Chapter 2. This will help us in defining some essential simplifications and assumptions that are needed to come to an appropriate DFB laser model. The main processes in a laser diode are the electrical carrier transport, carrier-photon interactions, optical wave propagation, and heat transport. Figure 3.1 shows how these processes interact with each other.

3.1.1 The Electrical Process: Electrical Carrier Transport

Through the electrical process, the carrier concentrations in the active layer are built up. The most essential part of this process is the carrier flow between the electrical contacts of the laser diode. This carrier flow is a combination of drift and diffusion processes across a multilayer structure. In (M)QW structures, tunneling and thermionic emission will also occur. The most important external factors that influence the carrier transport process are the electrical drive circuitry of the laser and the temperature.

Important electrical process variables are the carrier densities, the current density injected in the active layer, the current through the laser, and the voltage drop across the laser.

3.1.2 The Electro-Optic Process: The Carrier-Photon Interactions

The following carrier-photon interactions can be distinguished: stimulated emission, spontaneous emission, and absorption. The stimulated emission is characterized by the gain g and the associated variation in refractive index Δn_g. These parameters depend on the carrier concentrations, the photon density, the optical wavelength, and the temperature. They are discussed in Chapter 2. The gain and index parameters also affect the optical and electrical processes.

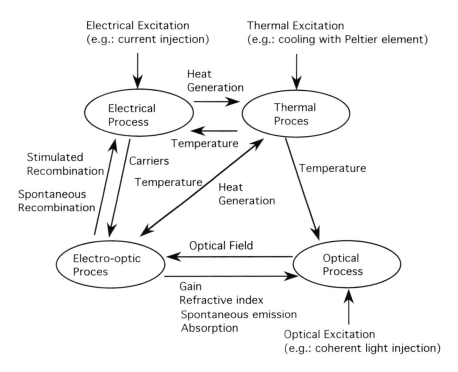

Figure 3.1 The main physical processes in a laser diode.

Important electro-optic process variables are the carrier densities in the active layer and the photon density of the different longitudinal modes.

3.1.3 The Optical Process: Optical Wave Propagation

Through the gain and spontaneous emission, an optical field can be established in the laser cavity. The power and the frequency of this field are determined by the phase and amplitude resonance conditions that the backward and forward propagating fields experience. Optical excitation corresponds to the injection of coherent light into one of the laser ends. Lasers operating below their threshold can be used to amplify the injected coherent light. Injection locking occurs if coherent light is injected in a laser operating above its threshold.

Important optical process variables are the phase, the amplitude, and the wavelength of the optical field associated with the different longitudinal modes. Potentially, the intensity and wavelength of an injected light signal may also be a relevant variable.

3.1.4 The Thermal Process: Heat Transport

The external heat sink temperature, the internal electrical and electro-optic heat dissipation, and the thermal (conductive and capacitive) properties of the laser structure and the heat sink determine the temperature of the active layer. This temperature influences the wavelength and the threshold current of the laser through its impact on the gain and the refractive index. Also, the lateral current confinement and transverse carrier confinement will be susceptible to changes in temperature.

Thermal excitation corresponds to controlling the temperature of the heat sink, for instance, with a Peltier element. For DFB lasers, Ti or Ti/Pt thin-film resistors may be integrated on the chip to heat the laser over a sufficiently large temperature range for wavelength fine-tuning.

Important thermal process variables are the temperatures of the active layer and of the heat sink.

In this book, thermal effects are not discussed. The active layer of the laser is assumed to be at room temperature ($T = 300K$) for any current level. This approach implies that calculated static laser characteristics (such as the P/I and λ/I curves) will not entirely fit experimental results. On the other hand, it must also be stressed that temperature variations hardly have an influence on the static side-mode suppression, the power spectrum, the intensity noise spectrum, or the modulation performance.

3.2 THE NEED FOR SIMPLIFICATION

A detailed description of a laser diode would require the determination of the carrier densities, the optical field, and the temperature at each point (x, y, z) of the laser. The detailed solution of this complex problem seems a far too ambitious, if not unattainable, goal. Furthermore, the physical insight and hence the derivation of easy design rules would not really benefit from such a complicated model. In the next sections, it is shown how a simplification can be brought about and approximations for the carrier density dynamics, for the field propagation, and for the interaction between field and carriers are introduced.

3.3 ASSUMPTIONS ABOUT THE MODELED LASER STRUCTURE

To determine the general outlook of the laser structure to be modeled, we go back to the basic requirements imposed by the telecommunications user. For telecommunications systems, three basic requirements can easily be stated. The laser needs to operate at low current levels, which implies a low threshold current. In this way the power

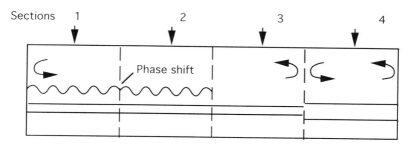

Figure 3.2 Longitudinal side view of a general multisection laser diode.

demands on the drive electronics are lowered and the lifetime expectancy is favored. The laser must also operate in a single lateral and transverse mode to obtain linear P/I curves and to have good coupling efficiencies with the waveguide or fiber-optic system to which the laser is attached. Furthermore, DSM operation or dynamic single-longitudinal-mode operation is required for long-distance fiber-optic transmission systems with high-speed modulation.

These performance requirements can be translated into the following structural requirements for the laser: strong lateral current confinement, strong lateral index guiding, and highly frequency selective feedback.

The latter requirement can be realized with cavities that contain gratings and/or that may be composed of several active or passive waveguide sections. Figure 3.2 illustrates a multisection DFB laser. Each section can be pumped independently and can have a different lateral/transverse geometry. An index- or gain-coupled grating can be present in the cladding layers or in the active layer. Furthermore, discrete reflections can occur at the interfaces between two sections or at the front and rear facets. The reflectivity at the interfaces depends on the discontinuities in the waveguide geometry between the sections, whereas the reflectivity at the facets is determined by the cleaving and/or coating of the facets.

The first two requirements can be realized with buried heterostructures. Figure 1.4(c) shows the schematic cross section of an etched-mesa buried heterostructure laser. Many other geometrical BH configurations can be found in the literature, but the lateral/transverse geometry always seems to have a similar degree of complexity. The active layer typically has a width w of 2 µm and a thickness d of 0.1 to 0.2 µm. No preference to a particular structure is forwarded here. Instead, it is assumed that for each section of a laser, the associated waveguide consists of a central cylindrical region with complex refractive index n_1 that is embedded in a surrounding index structure with refractive index $n_2(x, y)$. Figure 3.3 shows this generalized index profile of a laser diode waveguide.

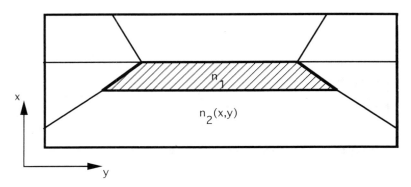

Figure 3.3 Generalized index profile of a BH laser diode.

3.4 OPTICAL WAVE PROPAGATION

From the previous discussion on physical processes, it is clear that the description of a DFB laser involves two types of basic variables: carrier densities and optical field amplitudes (when temperature variations are neglected). Therefore, two types of equations are also needed: carrier rate equations and longitudinal field equations. The latter are derived in this section.

3.4.1 Description of the Optical Field

The optical field is a vectorial quantity that depends on the spatial coordinates (x, y, z) and on time. The vectorial aspect corresponds to the polarization characteristics of the field. To obtain workable models, it is imperative that the optical field can be expressed in a more simplified way.

The time dependence of the field may be expressed in terms of several high-frequency components corresponding to different optical frequencies or laser modes. Moreover, on each of these high-frequency components, slow variations in time can be superimposed (for instance, in the form of modulation or noise).

The spatial dependence of the optical field may be described by a decomposition of the optical field into the basic modes of the laser waveguide. The (x, y)-dependence of the field is then expressed through the (x, y) field distribution of the modes of the waveguide. The z-dependence of the field can then be represented by z-dependent multipliers (or amplitude coefficients) appearing in each term of the decomposition. Moreover, it will be assumed further on that the lateral/transverse field distribution of the modes of the laser waveguide does not depend on time. This implies that the time dependence of the field must also be expressed through the multipliers.

Since each mode of the waveguide has a certain polarization, the decomposition of the field in modes of the waveguide also handles the description of the polarization of the laser field.

The lateral/transverse structure of the laser diode usually corresponds to a quasi-rectangular waveguide with quasi-TE and TM modes [1]. Since the lateral and transverse sizes of the active region are traditionally small (having a range of magnitude of the wavelength or less), only the lowest order TE and TM modes will be above cutoff [2]. In the remainder, only the lowest order TE mode is considered. The TM mode has higher mirror losses than the TE mode and is therefore suppressed by the lasing TE mode. This is also experimentally observed through polarization measurements on the laser output beam. This beam usually has a TE polarization, or in other words, the electrical field component mainly extends in the lateral y-direction.

For DFB lasers where the mirror losses at the facets may be less relevant due to the internal distributed reflection of the grating, this argument may have less value. In DFB lasers, the suppression of TM modes is nontrivial [3]. As will be discussed in Chapter 4, DFB lasers with a simple index-coupled grating and with no facet reflectivities show two longitudinal modes with equal feedback losses. One way to avoid this longitudinal mode degeneracy and to obtain DSM behavior is to have facet reflections. As already mentioned, the facet reflectivities are considerably higher for TE modes than for TM modes. Another way to eliminate the mode degeneracy can be realized by introducing a $\lambda/4$ phase shift in the grating of an index-coupled DFB laser with antireflection coatings on the facets. In this case, special attention must be paid to the suppression of the TM mode. To this end, the difference in coupling constants κ and the difference in the optical power confinement factors Γ for the TE and TM modes are exploited [4]. The parameters κ and Γ will be explained further on. TM modes are made to have smaller coupling constants than TE modes. In this way, the grating will reflect TM modes less well. Moreover, the overlap of the TE mode intensity distribution with the active region is usually better, corresponding to a higher optical confinement factor and leading to a higher modal gain for the TE mode.

The forward (+) and backward (−) propagating parts of the lateral electrical laser field of the lowest order TE mode in each waveguide section of a multisection laser are expanded as follows

$$E_y^\pm(x, y, z, t) = \mathrm{Re}\left\{\phi(x, y) \sum_m R_m^\pm(z, t) e^{j(\omega_m t \mp \beta_g z)}\right\} \quad (3.1)$$

The above field expansion relies on the assumption that the lateral/transverse field distribution of the electrical field is independent of the time t and the axial position z. This field distribution is represented by the distribution $\phi(x, y)$ of the lateral electrical field of the lowest order TE eigenmode of the waveguide. Furthermore, the expansion (3.1) assumes that the lateral electrical field distribution of the lowest order

TE eigenmode is identical for all longitudinal modes m because the optical frequencies of those different longitudinal modes are closely spaced.

The complex amplitudes $R_m^\pm$ in (3.1) represent those parts of the fields that vary slowly with time and with axial position. All rapid variations with time (on the order of the optical frequency) and with axial position (on the order of the internal wavelength) are included in the exponential factors. Although the choice of ω_m and β_g is not uniquely defined, ω_m and β_g can be considered as an angular reference frequency and a reference propagation constant, the choice of which must only lead to slowly varying amplitudes $R_m^\pm$. In what follows, ω_m will always be chosen so that it coincides with the optical frequency ν_m ($= \omega_m/2\pi$) of the mth longitudinal mode at its threshold. For β_g, the value $m\pi/\Lambda$ [$= (m/2)\beta_B$] is chosen if the section contains a grating of order m. Here, Λ denotes the period of the grating and $\beta_B = 2\pi/\Lambda$ the Bragg vector. In practical devices, Λ is usually chosen so that β_g nearly equals the propagation constant of the lowest order TE mode in the wavelength region for which the gain is maximum. When a section does not have a grating, β_g is set equal to the propagation constant β_m of the lowest order TE mode of the unperturbed waveguide (i.e., the waveguide without absorption or stimulated emission) at frequency ω_m.

3.4.2 The Scalar Wave Equation

Maxwell's equations provide the basic relationships that must be satisfied by the optical field. In semiconductor materials, such as GaAlAs and InGaAsP, which are isotropic with respect to light, they can be reduced to the wave equation [5]

$$\nabla^2 \mathbf{E} = \mu\varepsilon_0 \frac{\partial^2 \mathbf{E}}{\partial t^2} + \mu \frac{\partial^2 \mathbf{P}}{\partial t^2} \tag{3.2}$$

Here, $\mathbf{E}$ is the electrical field vector of the optical field. The polarization $\mathbf{P}$ in this equation consists of three distinctive parts: the polarization $\mathbf{P}_0$ of the structure without its grating, the polarization $\mathbf{P}_{pert}$ induced by a possible grating, and the polarization $\mathbf{P}_{spont}$ corresponding with the fluctuating spontaneous emission. The spontaneous emission causes phase and intensity noise in the modes and is introduced to have the linewidth and RIN characteristics covered by the model.

The first two polarizations $\mathbf{P}_0$ and $\mathbf{P}_{pert}$ can be expressed in terms of the corresponding parts of the refractive index:

$$\mathbf{P}_0 = \varepsilon_0 n^2(x, y, z, t)\mathbf{E} = \varepsilon_0[n_0(x, y) + \Delta n(x, y, z, t)]^2 \mathbf{E}$$

$$\mathbf{P}_{pert} = \varepsilon_0 n_p^2(x, y, z)\mathbf{E} \tag{3.3}$$

Here, $n_0(x, y, x)$ describes the refractive index structure of the laser waveguide without gain or losses and without grating. The perturbation in the refractive index Δn

represents the influence of the carrier density on the complex refractive index and includes contributions such as gain, absorption, and carrier-induced variations in the real part of the refractive index. The perturbation in the refractive index n_p represents the change in complex refractive index due to a grating. The grating is usually obtained by the introduction of a periodic thickness variation in one of the layers. Normally, Δn and n_p are small compared to n.

In order to replace the vectorial wave equation (3.2) with a simpler scalar wave equation, two approximations are applied. First, it is assumed that the time variations of Δn are slow compared to the optical frequencies. The time dependence of Δn becomes apparent in dynamic circumstances (e.g., current modulation) and is determined by the time dependence of the carrier density, which has a cutoff frequency of about 1 to 10 GHz. This is very low compared to the variations in time of the optical field, with optical frequencies on the order of 10^5 GHz.

Secondly, it is assumed that the lowest order TE mode corresponding with a lateral electrical field ($\mathbf{E} = E\mathbf{e_y}$) reaches its threshold first. Due to gain and carrier clamping (see Chapter 1), other lateral modes, if any, will therefore not be able to reach their threshold. This is the case for index-guided laser diode structures, which are commonly used for DFB laser diodes [6]. As such, only the scalar variable E of $E\mathbf{e_y}$ needs to be considered.

We can now replace the wave equation (3.2) with the scalar wave equation

$$\nabla^2 E - \mu\varepsilon_0 n^2 \frac{\partial^2 E}{\partial t^2} = F + \mu\varepsilon_0 n_p^2 \frac{\partial^2 E}{\partial t^2} \tag{3.4}$$

The term corresponding to the spontaneous emission polarization $\mathbf{P}_{spon}$ is now, for the sake of simplicity, denoted by F. The stochastic nature of spontaneous emission implies that F represents a stochastic driving force or a so-called Langevin function [7,8] for the wave equation.

3.4.3 The Langevin Force

It can be assumed that the Langevin force has a negligible spatial and temporal correlation, at least if we use F to denote the spontaneous emission, averaged over a small volume ΔV and during a short time Δt. Indeed, spontaneous emissions can only be correlated if they originate from the same carrier and if that carrier has not been scattered (and lost its phase information) in the time between the successive emissions. Δt and ΔV are thus determined by the typical scattering time ($\sim 10^{-13}$ sec) and the scattering distance ($\sim 10^{-2}$ μm).

The averaging over ΔV and Δt further implies that F can be approximated as a Gaussian process because it is the result of many processes. F is therefore completely characterized by its first- and second-order moments, which can be calculated quantum mechanically or semiclassically. A simple, semiclassical calculation has been reported in [9], where the moments of F are derived by requiring the spontaneously emitted light to be in equilibrium with the semiconductor. The moments are then found to be

$$\langle F(x, y, z, t)\rangle = 0$$

$$\langle F(x, y, z, t)F(x', y', z', t')\rangle = 2D_{FF}\delta(t - t')\delta(x - x')\delta(y - y')\delta(z - z')$$

$$2D_{FF} = \frac{4h\nu\,\omega^2}{\varepsilon_0 c^3}\,gnn_{sp} \tag{3.5}$$

Here g is the gain, n the refractive index, ω the angular optical frequency, and n_{sp} is given by

$$n_{sp} = \left[1 - \exp\left(\frac{h\nu - qV}{kT}\right)\right]^{-1} \tag{3.6}$$

3.4.4 Reduction Toward the Coupled-Wave Equations

The next step in simplifying the wave equations toward a workable, but still accurate, level of complexity is obtained by eliminating the (x, y)-dependence from the scalar wave equation (3.4). To this end, the expansion of the field $(E = E^+ + E^-)$ in a set of forward (E^+) and backward (E^-) propagating longitudinal modes, as proposed in (3.1), is used.

As already mentioned, the function $\phi(x, y)$ in (3.1) represents the field distribution of the lowest order TE mode of the waveguide described by the index profile $n = n_0 + \Delta n$ (no grating). For each longitudinal mode m with angular reference frequency ω_m, $\phi(x, y)$ obeys the two-dimensional Helmholtz equation

$$\nabla^2_{xy}\phi + \omega_m^2\mu_0\varepsilon_0 n^2\phi = \beta^2_{c,m}\phi \tag{3.7}$$

Remember that ϕ is assumed to be identical for all longitudinal modes m and to be independent of the longitudinal coordinate z and the time coordinate t. This implies that $\beta_{c,m}$ is z-, t-, and m-dependent. In practice, the field distribution $\phi(x, y)$ can be approximated by the field distribution $\phi_0(x, y)$ of the lowest order TE eigenmode of the unperturbed waveguide with refractive index profile $n_0(x, y)$ and propagation constant β_m for the longitudinal mode m. The complex eigenvalue $\beta_{c,m}$ can then be expressed as a function of β_m, $\Delta n(x, y, z, t) = n - n_0$, ϕ_0 and ω_m as

$$\beta_{c,m} - \beta_m = \frac{\omega_m^2}{c^2 \beta_m} \frac{\iint_{x,y} n_0(n - n_0)\phi_0^2 \, dx \, dy}{\iint_{x,y} \phi_0^2 \, dx \, dy} \qquad (3.8)$$

Several techniques exist to calculate ϕ_0 and β_m. The latter will often be expressed in terms of the effective refractive index n_e through the relation $\beta_m = 2\pi n_e/\lambda_m$, n_e being assumed m-independent. If n and n_0 do not vary within the (x, y) cross sections of the layers, (3.8) can be approximated by

$$\beta_{c,m} = \frac{2\pi}{\lambda_m} \left\{ n_e + \sum_i (n_i - n_{oi}) \frac{n_{oi}}{n_e} \Gamma_i \right\}$$

$$\Gamma = \frac{\iint_{\text{layer } i} \phi_0^2(x, y) \, dx \, dy}{\iint_{(x,y)-\text{plane}} \phi_0^2(x, y) \, dx \, dy} \qquad (3.9)$$

The factor Γ_i is the confinement or power filling factor of the ith layer and n_{0i} is the refractive index of the ith layer in the unperturbed situation (without gain, losses, or grating).

When the cross section of the laser is further simplified to an active layer and a set of two surrounding cladding layers, $\beta_{c,m}$ can be written as

$$\beta_{c,m} = \frac{2\pi}{\lambda_m} (n_e + \Gamma \Delta n_r) + j0.5 \, \Gamma g(\lambda_m) - j0.5 \, \alpha_{\text{int}} \qquad (3.10)$$

with $\alpha_{\text{int}} = \Gamma \alpha_{ac} + (1 - \Gamma)\alpha_{cl}$ and with Γ the confinement or power-filling factor of the active layer. Furthermore, g, α_{ac}, and Δn_r are, respectively, the gain, the absorption, and the related changes in real refractive index in the active layer, while α_{cl} represents the absorption in the cladding layers.

If the scalar field E in the scalar wave equation (3.4) is now substituted with the expansion (3.1), then the lateral and transverse dependence expressed through $\phi(x, y)$ can be eliminated using the Helmholtz equation (3.7). Also, the fast temporal and spatial variations of the optical field can be eliminated from the scalar wave equation through the slowly varying amplitude approximation of expansion (3.1) in which the fast-varying parts have been made explicit via the factors $\exp(j\omega_m t)$ and $\exp(\pm j\beta_g z)$ [10]. Therefore, after $\phi(x, y)$ is eliminated, the resulting wave equation is averaged over a few time periods T_m (with T_m being defined by the mode spacing $\delta\omega_m = 2\pi/T_m$) and over a few grating periods (or if no grating is present, over a few wavelength periods). Averaging over a few periods T_m allows the scalar wave equation to be decomposed in separate longitudinal wave equations for each longitudinal mode considered in the expansion (3.1). Averaging over a few grating periods or wavelength periods allows the longitudinal wave equation for each mode to be further decomposed into a set of two coupled-wave equations describing, respectively, the forward and backward propagating waves. These coupled-wave equations are

$$\frac{\partial R_m^+}{\partial z} + \frac{1}{v_g}\frac{\partial R_m^+}{\partial t} + j\Delta\beta_m R_m^+ = \kappa_{FB} R_m^- + F_m^+$$

$$-\frac{\partial R_m^-}{\partial z} + \frac{1}{v_g}\frac{\partial R_m^-}{\partial t} + j\Delta\beta_m R_m^- = \kappa_{BF} R_m^+ + F_m^- \qquad (3.11)$$

The coupling coefficients κ_{FB} and κ_{BF}, the group velocity v_g, and the complex Bragg deviation $\Delta\beta_m$ in (3.11) are given by the following expressions

$$\kappa_{FB} = -j\frac{\omega_m^2 a_{-n}}{2\beta_g c^2}, \quad \kappa_{BF} = -j\frac{\omega_m^2 a_n}{2\beta_g c^2} \qquad (3.12a)$$

$$n_{p,\text{eff}}^2 = \sum_{q=-\infty}^{+\infty} a_q e^{j\frac{2q\pi}{\Lambda}z} \qquad (3.12b)$$

$$n_{p,\text{eff}}^2 = \frac{\iint_{x,y} n_p^2 \phi_0^2 dx\, dy}{\iint_{x,y} \phi_0^2 dx\, dy} \qquad (3.12c)$$

$$\frac{\beta_m}{v_g} = \frac{\omega_m}{c^2} = \frac{\iint_{x,y} n^2 \phi_0^2 \mathrm{d}x\,\mathrm{d}y}{\iint_{x,y} \phi_0^2 \mathrm{d}x\,\mathrm{d}y} \tag{3.12d}$$

$$\Delta\beta_m = \beta_{c,m} - \beta_g \tag{3.12e}$$

The group velocity v_g in (3.12d) is considered as a constant with the same value for all longitudinal modes. The function $n_{p,\mathrm{eff}}^2(z)$ defined in (3.12) corresponds to the effective index variation caused by the refractive index perturbation $n_p(x, y, z)$ of the grating. It represents a perturbation of n_e and because of its periodic z-dependence, it can easily be expanded in a Fourier series with period Λ. As shown in (3.12), the Fourier series consists of an infinite number of harmonic components with their coefficients denoted by a_q ($-\infty < q < +\infty$). These harmonic components can cause strong interactions between the forward and backward propagating guided waves of the laser waveguide if they fulfill the Bragg condition

$$\beta_{c,m} \approx q(2\pi/\Lambda) - \beta_{c,m} = q\beta_B - \beta_{c,m} \tag{3.13}$$

If the nth Fourier component approximately fulfills this condition, we speak about an nth-order grating or DFB laser. For an nth-order DFB laser, the grating must be designed toward high values for the coefficients a_n and a_{-n}, because, as (3.12a) shows, these coefficients determine the coupling coefficients κ_{FB} and κ_{BF}.

For a pure index grating, in other words, for a periodic variation of the real refractive index, the Fourier coefficients of the real function $n_{p,\mathrm{eff}}^2(z)$ obey the relation $a_n = a_{-n}^*$ and hence $\kappa_{BF} = -\kappa_{FB}^*$. A pure gain grating corresponds to an imaginary $n_{p,\mathrm{eff}}^2$ and $\kappa_{BF} = \kappa_{FB}^*$. For complex coupling, κ_{FB} and κ_{BF} are often represented by, respectively, $\kappa_{FB} = \kappa_g + j\kappa_i$ and $\kappa_{BF} = \kappa_g^* + j\kappa_i^*$, where κ_g represents the gain coupling and k_i the index coupling. Note that the phases of κ_{BF} and κ_{FB} depend on the choice of the origin along the z-axis ($z = 0$) relative to the grating.

The Fourier expansion can also be applied to quasiperiodic gratings, that is, gratings whose period (chirped gratings) or amplitude (pitch modulated gratings) vary slowly in the longitudinal direction. In this case, the Fourier coefficients a_q are no longer constant and their phase and amplitude become z-dependent. Therefore, in the most general case, the coupling coefficients κ_{FB} and κ_{BF} can be slowly varying functions of z. Examples of such cases will be shown in Chapter 6. We will treat the coupling coefficients in the following as given constants.

For an nth-order grating, β_g is chosen equal to $\beta_g = n\pi/\Lambda$. The Bragg deviation $\Delta\beta_m$ indicates then how well $\beta_{c,m}$ fulfills the Bragg condition (3.13) for $q = n$. If no grating is present, β_g is chosen equal to β_m, with m corresponding to the strongest lasing mode. In this case, the spatial averaging that leads to (3.11) is done over a few wavelength periods and the coupling coefficients κ_{FB} and κ_{BF} in (3.11) will vanish. The latter case describes the longitudinal wave propagation in a Fabry-Perot laser diode.

Since the mode spacing $\delta\omega_m$ is typically on the order of 100 GHz, the coupled-wave equations (3.11) are only valid for modulation frequencies up to a few times 10 GHz. The field quantities $R_m^\pm(z, t)$, therefore, denote values averaged over a few times 10 ps.

The $F_m^\pm$ terms in (3.11) represent the Langevin forces corresponding to the spontaneous emission. They can be obtained from the Langevin function F by purely linear transformations. Therefore, their first-order moments are zero and the second-order moments can easily be calculated [5]

$$\langle F_l^\pm(z, t)\, F_m^\pm(z', t')\rangle = 0$$

$$\langle F_l^\pm(z, t)\, F_m^{\pm *}(z', t')\rangle = h\nu_m n_{sp}\Gamma g(\omega_m)\delta(z - z')\delta(t - t')\delta_{lm}$$

$$\langle F_l^-(z, t)\, F_m^{+ *}(z', t')\rangle = \langle F_l^+(z, t)\, F_m^{- *}(z', t')\rangle$$

$$= 2h\nu_m n_{sp} \frac{\kappa_{FB} + \kappa_{BF}^*}{2} \delta(z - z')\delta(t - t')\delta_{lm} \tag{3.14}$$

The last moment of (3.14) is only nonzero in case of gain gratings with corrugated active layers (i.e., not for loss gratings in the cladding layers). In reality, the forward and backward propagating spontaneous emissions are not correlated. However, F_m^+ and F_m^- represent the spontaneous emission averaged over a distance that is large compared to the grating period and only these averages are correlated.

Note that these moments depend on the choice of the field normalization. The normalization of ϕ is chosen such that the average optical power in the $+$ or $-$ direction can be expressed as

$$P^\pm = \sum_m P_m^\pm = \sum_m |R_m^\pm|^2 \tag{3.15}$$

Equation (3.15) implies the following normalization

$$\iint_{(x,y)} |\phi(x,y)|^2 \, dx \, dy = \frac{2\mu_0 \omega}{\beta_g} \approx \frac{2\mu_0 c}{n_e} \qquad (3.16)$$

The energy density (per unit distance in the longitudinal direction) inside the laser cavity can then be written as [11]

$$u = \frac{1}{v_g} \sum_m (|R_m^+|^2 + |R_m^-|^2) \qquad (3.17)$$

This energy density is obviously connected with the photon density inside the cavity. From (3.15) and (3.17), it can be concluded that the number of photons $s_m^\pm$ in mode m per unit length in the longitudinal direction and propagating in the $\pm$ direction is given by

$$s_m^\pm = |R_m^\pm|^2 / (v_g h \nu_m) \qquad (3.18)$$

For a detailed derivation of the coupled-wave equations (3.11), the reader should see [5]. The derivation of the coupled-wave equations can also be found in many other textbooks on optics [2,12,13]. However, these descriptions usually do not consider slow variations in time, noise sources, and contributions from spontaneous emission. Moreover, they do not take into account any dependencies of the complex refractive index on the optical power (via the stimulated recombination).

3.5 DISCUSSION OF THE COUPLED-MODE WAVE EQUATIONS

3.5.1 The Bragg and Maximum Gain Wavelengths

In general, for each wavelength range in the vicinity of a Bragg wavelength $\lambda_{B,q} = 2n_e \Lambda / q$ ($q = 1, 2, \ldots$), the Bragg condition (3.13) will hold, which means that a strong interaction occurs between the counterpropagating guided waves at the same wavelength. With each interaction there is a corresponding Fourier component of the Fourier expansion of the periodic (complex) index variation.

If the grating structure is considered to be a Bragg reflector (as in Figure 1.20), a Bragg reflection characteristic can be drawn. Figure 3.4 shows a series of Bragg windows corresponding to different Fourier components. The height and width of such a Bragg window depend on the strength of the associated Fourier coefficients and on the

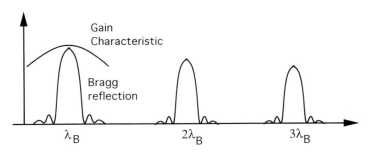

Figure 3.4 Series of wavelength windows with a strong Bragg reflection. For a first-order grating, the first window is positioned in the wavelength region with maximum gain.

length of the grating. For $q = 1$ and $2n_e\Lambda \approx 1.55$ µm, the width is usually only a few nanometers.

The difference in Bragg wavelengths $\Delta\lambda_{B,q} = \lambda_{B,q} - \lambda_{B,q+1}$ for two adjacent Bragg windows is given by $\Delta\lambda_{B,q} = \lambda_{B,q}/(q + 1)$. For $q = 1$, $\Delta\lambda_{B,q} \approx 0.75$ µm in 1.55-µm DFB lasers. Therefore, the other Bragg windows will fall outside the gain region if the grating period Λ is optimized to have one Bragg window centrally positioned in the maximum gain region. Figure 3.4 illustrates this. Consequently, the expansion of (3.1) only needs to consider longitudinal modes with wavelengths around the gain maximum and around only one Bragg window.

If a DFB laser is badly designed and the gain maximum does not coincide well enough with the Bragg window, then the laser can show DFB lasing modes as well as Fabry-Perot lasing modes. This assumes of course a DFB laser with no antireflection coatings on its facets. Figure 3.5 illustrates this design issue.

3.5.2 Influence of Radiation Modes in Higher Order Gratings

We started our discussion of optical wave propagation by stating that the waveguide of the laser is usually designed in such a way that only the lowest order quasi-TE and TM guided modes are below their cutoff wavelength at the lasing wavelength. Methods to suppress the TM mode are discussed in Section 3.4.1. However, the laser waveguide also has a continuum of radiation (evanescent) modes below cutoff. Guided waves show an exponential decay in the cladding layers, and therefore the energy guided by these waves is confined to the vicinity of the guiding layer. Radiation modes vary sinusoidally in at least one of the cladding regions. Consequently, if the field varies, for instance, sinusoidally in the substrate, power will be radiated in the substrate. Radiation modes have a much higher threshold gain than the guided wave. They suffer from radiation losses, and, in addition, their overlap with the spatial gain distribution is worse than for the guided wave. In Fabry-Perot lasers, radiation modes do not occur except near the laser facets due to some power redistribution across the modes by the facet reflection [14].

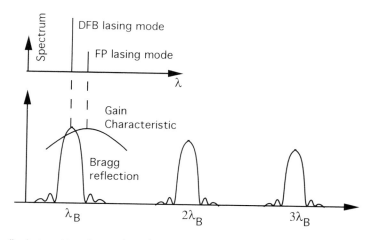

Figure 3.5 Badly designed DFB laser, where the maximum gain region is ill-positioned with respect to the Bragg wavelength.

The presence of the grating, however, allows coupling between the two counter-propagating guided waves (both lowest order quasi-TE) and radiation modes. For reasons of simplicity we illustrate this by means of a basic thin-film waveguide with one corrugated guiding layer (with refractive index n_1) and two infinite claddings (with refractive indexes $n_2, n_3 < n_1$). The waveguide is assumed to support only the lowest order guided TE mode. A guided wave traveling in the $+z$-direction with propagation constant $\beta_{c,m}$ will interact with the Fourier components of the grating to excite modes with propagation constants $\beta_r \approx \text{Re}(\beta_{c,m}) - 2\pi p/\Lambda$ ($p = \pm 1, \pm 2, \ldots$). Radiation modes with propagation constant β_r only exist if $|\beta_r| < 2\pi n_i/\lambda_m$ ($i = 2, 3$), where λ_m is the free space wavelength of the considered light. Therefore, in a first-order grating with $\text{Re}(\beta_{c,m}) \approx \pi/\Lambda$ and $2\pi n_i/\lambda_m < \text{Re}(\beta_{c,m}) < 2\pi n_1/\lambda_m$ ($i = 2, 3$), no radiation modes can be excited. However, for higher order gratings, this is no longer the case. Figure 3.6 shows the relation between the propagation constants of the interacting guided and evanescent modes for the simple thin-film waveguide [15]. Examples for second-order and third-order gratings are given. For a radiation mode with propagation constant β_r, the radiation angle in the ith cladding layer ($i = 2, 3$) is given by $\theta_{r,i} = \arccos[\beta_r/(2\pi n_i/\lambda_m)]$ [16]. For a second-order grating, $\text{Re}(\beta_{c,m}) \approx 2\pi/\Lambda = \beta_B$. This means that for $p = \pm 1$ the guided mode propagating in the $\pm z$ direction will couple with radiation modes for which $\beta_r \approx 0$ holds. In this case, the radiation angle is 90° and first-order radiation losses will occur.

When radiation modes are excited by the guided mode, these radiation modes will in turn also excite the guided mode. To incorporate these secondary coupling effects, the coupled-mode wave equations need to be extended as follows

1st order grating

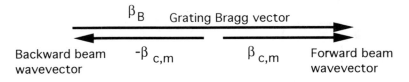

2nd order grating

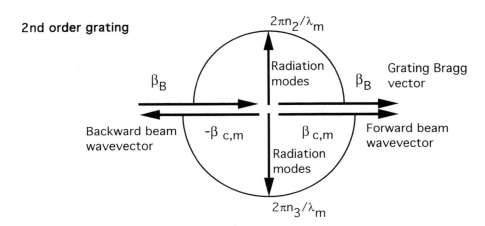

3rd order grating

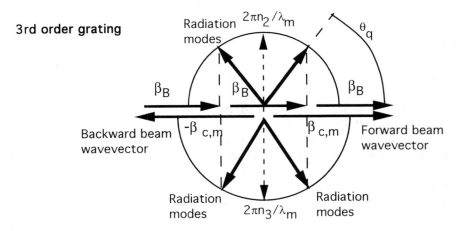

Figure 3.6 Graphical representation of the relation between the wave vectors of the guided mode, the radiation modes, and the grating for a simple thin-film waveguide. (*After:* [15].)

$$-\frac{\partial R_m^+}{\partial z} + \frac{1}{v_g}\frac{\partial R_m^+}{\partial t} + j(\Delta\beta_m + \zeta^+)R_m^+ = F_m^+ + (\kappa_{FB} + \kappa_r^+)R_m^-$$

$$-\frac{\partial R_m^-}{\partial z} + \frac{1}{v_g}\frac{\partial R_m^-}{\partial t} + j(\Delta\beta_m + \zeta^-)R_m^- = F_m^- + (\kappa_{BF} + \kappa_r^-)R_m^+ \quad (3.19)$$

The terms $\zeta^{\pm}$ and $\kappa_r^{\pm}$ are corrections due to the radiation effects. They are only relevant for higher order gratings. Their derivation is rather cumbersome, and for a detailed description the reader can refer to [15,17–19]. We limit ourselves here to an interpretation of these corrective terms. To focus the discussion, we only look at ζ^+ and κ_r^+.

The ζ^+ coefficient represents the effect of the forward guided wave exciting radiation modes (for which $\beta_r \approx \mathrm{Re}(\beta_{c,m}) - 2\pi p/\Lambda$ holds) and where those radiation modes in turn act on the forward guided wave via the opposite Fourier component $2\pi p/\Lambda$. The imaginary part of ζ^+ describes radiation losses, and therefore its value is negative. The real part of ζ^+ causes an extra detuning. One can show that for symmetrically or asymmetrically shaped index gratings, $\zeta^+ = \zeta^-$ [17].

The κ_r^+ coefficient represents the effect of the forward guided wave exciting radiation modes (for which $\beta_r \approx \mathrm{Re}(\beta_{c,m}) - 2\pi p/\Lambda$ holds) and where those radiation modes in turn act on the backward guided wave via the Fourier component $2\pi(p - n)/\Lambda$ (assuming an nth-order grating). Therefore, κ_r^+ is a correction to κ_{FB}. For symmetrically shaped index gratings, one can show that $\kappa_r^+ = (\kappa_{FB}/\kappa_{BF})\kappa_r^-$ [17]. In general κ_r^+ and κ_r^- are complex, and they will cause an overall gain/loss coupling (even in an index-coupled grating).

The special case of DFB lasers with a symmetrically shaped second-order index grating is discussed in the next chapter.

At present, first-order gratings are used in most DFB lasers. However, second-order gratings are sometimes preferred because they are usually easier to make, since their grating period is twice as large as that of first-order gratings. In particular, in the early period of DFB lasers when the technology for the development of gratings was less mature, second-order gratings were popular. Also, some attempts were undertaken to obtain surface-emitting DFB (or DBR) lasers.

In the remainder of this book we will omit the corrections $\zeta^{\pm}$ and $\kappa_r^{\pm}$ unless we explicitly deal with higher order DFB lasers.

3.5.3 Longitudinal Rate Equations for the Optical Field

A better understanding of the coupled-wave equations (3.11) can be obtained by converting these equations into longitudinal rate equations describing the behavior of the amplitudes and phases of the complex field variables $R_m^\pm$. To this end, we first characterize the complex quantities $R_m^\pm$, $F_m^\pm$ by their amplitude and phase. We therefore write

$$R_m^\pm(z, t) = r_m^\pm(z, t)\exp[j\varphi_m^\pm(z, t)]$$

$$F_m^\pm(z, t) = |F_m^\pm(z, t)|\exp[j\varphi_{F,m}^\pm(z, t)]$$

$$\Delta\beta_m(z, t) = \Delta\beta_{m,r}(z, t) + j\Delta\beta_{m,i}(z, t) \quad (3.20)$$

$$\kappa_{FB}(z) = |\kappa|\exp(j\varphi_\kappa)$$

Substitution of (3.20) in the coupled-wave equations (3.11) results in the longitudinal rate equations for the amplitudes and phases of $R_m^\pm$:

$$\pm\frac{\partial r_m^\pm}{\partial z} + \frac{1}{v_g}\frac{\partial r_m^\pm}{\partial t} - \Delta\beta_{m,i} r_m^\pm = |F_m^\pm|\cos(\varphi_{F,m}^\pm - \varphi_m^\pm) \pm |\kappa| r_m^\mp \cos(\varphi_\kappa + \varphi_m^- - \varphi_m^+)$$

$$(3.21)$$

$$\pm\frac{\partial\varphi_m^\pm}{\partial z} + \frac{1}{v_g}\frac{\partial\varphi_m^\pm}{\partial t} + \Delta\beta_{r,m} = \frac{|F_m^\pm|}{r_m^\pm}\sin(\varphi_{F,m}^\pm - \varphi_m^\pm) + |\kappa|\frac{r_m^\mp}{r_m^\pm}\sin(\varphi_\kappa + \varphi_m^- - \varphi_m^+)$$

$$(3.22)$$

The last term on the right-hand side of (3.21) corresponds to the reflected power of the backward (forward) propagating wave that couples into the forward (backward) propagating wave. The reflected backward (forward) propagating fields are not necessarily in phase with the forward (backward) propagating fields. The phase mismatch, expressed by $\varphi_\kappa + \varphi_m^- - \varphi_m^+$ in the coupling term of (3.21), results in an imperfect energy (photon) transfer from the reflected waves. The interference effect also affects the phases of forward and backward propagating waves, as can be seen from (3.22). The influence of the interference obviously depends on the relative strength of reflected and transmitted waves; for example, a weak reflected wave r_m^- will only have a small impact (expressed by r_m^-/r_m^+) on the phase of the resulting field.

The phase variation due to the Bragg deviation accounts for the discrepancy between the actual wave vector and the chosen reference Bragg wave vector when a grating is present. In this case, the Bragg deviation determines the phase mismatch and hence the efficiency of the distributed feedback.

3.5.4 The Instantaneous Optical Frequencies

With the phases given by (3.22), a set of forward and backward instantaneous optical frequencies $\varpi_m^\pm(z, t)$ can be defined for each position along the laser cavity

$$\varpi_m^\pm(z, t) = \omega_m + \frac{\partial \varphi_m^\pm(z, t)}{\partial t} \qquad (3.23)$$

The index m refers to the different longitudinal modes. If the dynamic excitation is slow compared to the round-trip time of the laser (a few picoseconds), then the instantaneous frequencies will hardly vary along the cavity. Therefore, the instantaneous frequency ϖ_m^- or ϖ_m^- at any position along the laser can be used to represent the dynamically varying instantaneous lasing frequency $\varpi_m(t)$ of the mth mode. Here, the instantaneous frequency of the backward propagating field at the left-hand facet is taken. If the origin of the z-axis is chosen at this facet, ϖ_m becomes $\varpi_m^-(z = 0, t) = \omega_m + \partial \varphi_m^-(0, t)/\partial t$. For calculation purposes, it is often useful to introduce the variable $\Delta\omega_m(t) = \varpi_m - \omega_m$, as already used in Chapter 2 for the rate equation model.

Finally, it can be noted that the slowly varying amplitude approximations used to derive the wave equations (3.11) imply the conditions $\Delta\omega_m \ll \omega_m$ and $|\Delta\beta_m| \ll \beta_g$.

3.5.5 Comments on Spontaneous Emission

In this section we will further refine the interpretation of the wave equations and the associated longitudinal rate equations with respect to the contribution of spontaneous emission to the average power.

By averaging (3.21), it can be shown that spontaneous emission does not contribute to the average amplitudes $\langle r_m^\pm \rangle$ because the Langevin term in (3.21) has a zero average. However, for the average power or the number of photons per unit length, different results appear. Multiplication of the appropriate equation in (3.11) with $(R_m^+)^*$ or $(R_m^-)^*$, adding the complex conjugate of the resulting expression and using (3.18), results in the following equation for the number of photons $s_m^\pm$ per unit length along the cavity in the mth mode:

$$\frac{\mathrm{d}s_m^\pm}{\mathrm{d}t_{f,b}} - 2\Delta\beta_{i,m} v_g s_m^\pm = \frac{R_{sp,m}}{2L} \pm 2|\kappa| v_g \sqrt{s_m^+ s_m^-} \cos(\varphi_\kappa + \varphi_m^- - \varphi_m^+) + F_{sm}^\pm$$

$$F_{sm}^\pm(z, t) = \frac{R_m^{\pm *} F_m^\pm + R_m^\pm F_m^{\pm *} - \langle R_m^{\pm *} F_m^\pm + R_m^\pm F_m^{\pm *} \rangle}{h\nu_m} \qquad (3.24)$$

with $d/dt_{f,b} = \partial/\partial t \pm v_g \partial/\partial z$ and $R_{sp,m} = \Gamma g(\lambda_m) n_{sp} v_g$ the rate of spontaneous emission that couples into mode m [20]. $R_{sp,m}$ is a fraction of the total spontaneous emission rate $BN^2 V_{act}$, with V_{act} the volume of the active layer and BN^2 the number of photons generated spontaneously per unit time and per unit volume (i.e., the radiative bimolecular recombination). Therefore, alternatively, $R_{sp,m}$ can also be written as $R_{sp,m} = \beta_{sp}(\lambda_m) BN^2 V_{act}$, with β_{sp} the spontaneous emission factor as defined in [21]. The new Langevin functions $F_{sm}^{\pm}$ have a zero mean. Their second-order moments can easily be calculated and are given in [5].

Equation (3.24) indicates that the spontaneous emission contributes to the stochastically averaged power of the modes. Moreover, the spontaneous emission rate $R_{sp,m}/L$ that couples into the mth longitudinal mode per unit distance in the longitudinal direction is divided equally between the forward and backward propagating waves. This corresponds to the fact that the spontaneously emitted photons have uniformly distributed propagation directions in space.

We thus observe that the noise does not contribute to $\langle r_m^{\pm} \rangle$, but does contribute to $\langle (r_m^{\pm})^2 \rangle$. This is due to the random phase of the spontaneously emitted photons (see also Figure 2.2) and basically relates to the fact that $\langle (r_m^{\pm})^2 \rangle \neq \langle r_m^{\pm} \rangle^2$. This peculiarity must be treated with care in order to obtain correct calculation results.

The coupled-wave equations (3.11) are easier to work with for continuous wave (CW) or dynamic analysis if the Langevin forces are omitted. This seems acceptable at first glance. However, this would mean that for $r_m^{\pm} = \langle r_m^{\pm} \rangle + \Delta r_m^{\pm}$ (and $\langle \Delta r_m^{\pm} \rangle = 0$), the nonzero average $\langle (\Delta r_m^{\pm})^2 \rangle$ must be neglected in the intensity equation. For small fields (i.e., close to the threshold of the lasing mode or for side modes) and also in the case of a small-signal analysis, this omission causes nonneglectable errors. Therefore, if in field calculations the Langevin forces are neglected or if in a small-signal approximation the Langevin contributions are approximated by linear terms in the amplitude and phase noise, a new complex amplitude $A_m^{\pm}$ should be used for which $\langle |A_m^{\pm}| \rangle^2 = \langle |A_m^{\pm}|^2 \rangle = \langle (r_m^{\pm})^2 \rangle$ holds. This also means that $\langle (\Delta a_m^{\pm})^2 \rangle = 0$, with $|A_m^{\pm}| = a_m^{\pm} = \langle a_m^{\pm} \rangle + \Delta a_m^{\pm}$ (and $\langle \Delta a_m^{\pm} \rangle = 0$), whereas $\langle (\Delta r_m^{\pm})^2 \rangle \neq 0$.

Substitution of the expansion,

$$(r_m^{\pm})^2 = \langle a_m^{\pm} \rangle^2 + 2\langle a_m^{\pm} \rangle \Delta a_m^{\pm} + (\Delta a_m^{\pm})^2 \tag{3.25}$$

in (3.24) followed by linearization in $\Delta a_m^{\pm}$ and statistical averaging, leads to two equations, one for $\Delta a_m^{\pm}$ and one for $\langle a_m^{\pm} \rangle$. Adding those two equations then gives the equation for $a_m^{\pm}$. The phase of $A_m^{\pm}$ can still be described by the phase $\varphi_m^{\pm}$ of $R_m^{\pm}$ and by (3.22). Combining now the equations for $a_m^{\pm}$ and $\varphi_m^{\pm}$ finally gives a set of coupled-wave equations for $A_m^{\pm}$

$$\frac{\partial A_m^+}{\partial z} + \frac{1}{v_g}\frac{\partial A_m^+}{\partial t} + \left\{ j\Delta\beta_m - \frac{R_{sp,m}}{4L|A_m^+|^2} \right\} A_m^+ = \kappa_{FB} A_m^-$$

$$-\frac{\partial A_m^-}{\partial z} + \frac{1}{v_g}\frac{\partial A_m^-}{\partial t} + \left\{ j\Delta\beta_m - \frac{R_{sp,m}}{4L|A_m^-|^2} \right\} A_m^- = \kappa_{FB} A_m^+ \qquad (3.26)$$

Our computer simulations are usually based on (3.26). For noise analysis they also need to be extended with appropriate Langevin forces. A detailed derivation of (3.26), including the Langevin forces, can be found in [5]. A more intuitive derivation of (3.26) is given in [22]. Note that spontaneous emission is written here as a contribution to the gain.

From a notational point of view, we will use $R_m^\pm$ in the remainder of this book, even if for calculation purposes the amplitudes $A_m^\pm$ are used.

3.6 THE ELECTRICAL TRANSPORT PROBLEM

A classical macroscopic description of the carrier flow in the semiconductor structure of a laser diode would involve several elements: the Poisson equation describing the electrical potential distribution, the carrier continuity equations, the drift and diffusion current equations, the Fermi-Dirac statistics for the energy distribution of the carriers, the valence and conduction band structures (usually assumed parabolic), and, finally, the constitutive laws between the electrical drift field and the displacement field.

3.6.1 The Carrier Rate Equation

For our purpose, the above relations are far too complex to work with. However, in most cases this problem can be simplified because the basic variables of interest are the carrier density in the active layer and the current injected in the laser. Therefore, it is assumed that all laser sections are current-controlled and that due to leakage currents only a fraction η of the injected carriers reaches the active region of the BH laser, where they are distributed uniformly in the lateral and transverse direction. This assumption is based on the observation that the active layer width (1 to 3 µm) and thickness (0.1 to 0.2 µm) are usually smaller than the diffusion length of the carriers (>3–5 µm). Furthermore, no doping is assumed in the active layer and hence it follows from the neutrality condition that the hole density P equals the electron density N. The following rate equation for the carrier density $N(z, t)$ in the active layer then holds

$$\frac{\partial N}{\partial t} = \frac{\eta J}{qd} - AN - BN^2 - CN^3 - \frac{\Gamma}{wd}\sum_m \frac{g(\omega_m)}{h\nu_m}[(r_m^+)^2 + (r_m^-)^2] + F_N(z, t) \quad (3.27)$$

This rate equation is similar to (2.2), except that here the carrier density relates to a longitudinal position along the laser and that optical field powers are used instead of photon densities. The second to last term in (3.27) expresses the stimulated recombination.

It must be noted that in (3.27) the carrier diffusion in the active layer is neglected in the lateral/transverse and longitudinal directions. Omission of the longitudinal diffusion is justified by the fact that the carrier diffusion length is generally small with respect to the typical distance over which the carrier density varies due to the other terms in (3.27). Lateral/transverse diffusion is neglected because the lateral/transverse dimensions of the active layer are usually small compared to the diffusion length (3 to 5 µm).

Sometimes, for wider active layers, the lateral diffusion may be relevant. This case can best be treated by developing the carrier density $N(y, z, t)$ and the lateral optical field intensity distribution in a Fourier series across the active layer width w [23]. A first-order approximation for N is then given by $N(y, z, t) = N_a(z, t) - N_s(z, t) \cos(2\pi y/w)$, leading to two carrier rate equations, respectively, for N_a and for N_s. This approach will not be elaborated upon any further because DFB lasers usually show sufficiently narrow active layers.

The rate equation (3.27) also implies that in a multisection laser, current leakage from one longitudinal section to another is neglected, which is only correct if all sections are isolated electrically.

In the carrier rate equation, we have treated the carrier density as a continuous quantity, while the recombination and creation (e.g., interband absorption) rates have approximated as continuous processes. However, in reality, the carrier density is quantized with a unit equal to the inverse active layer volume, and the recombination and creation processes occur in a discrete way. This discrete character can be accounted for by the Langevin force $F_N(z, t)$ in (3.27).

For the same reasons as those given in Section 3.4.3, we can assume that $F_N(z, t)$ and $F_N(z', t')$ are Gaussian and uncorrelated for $|t - t'| > 10^{-13}$ sec and $|z - z'| > 10^{-2}$ µm. Again, F_N has a zero average. A part of F_N has its origin in the spontaneous emission and is thus related to the Langevin functions we encountered in the previous section. However, spontaneous emissions induce opposite changes in photon number and in carrier number (one spontaneous emission increases the photon number by 1, but it decreases the carrier number by 1). We can therefore write

$$wd\, F_N(z, t) = wd\, F_S(z, t) - \sum_m \{F_{sm}^+(z, t) + F_{sm}^-(z, t)\} \tag{3.28}$$

The multiplication with wd is necessary to obtain the number of carriers per unit length in the longitudinal direction, a quantity similar to $s_m^\pm$. Note that $F_m^\pm$ can be interpreted as shot noise associated with the stimulated emission and the interband absorption [9].

Furthermore, wdF_S represents the shot noise related to the spontaneous carrier recombination. This shot noise is not correlated with the shot noise in stimulated emission and absorption (independent processes) and its second-order moment can be derived from the standard formula for shot noise. One finds

$$\langle F_S(z, t)F_S(z', t') \rangle = \frac{2}{wd}(AN + BN^2 + CN^3)\delta(t - t')\delta(z - z') \quad (3.29)$$

3.6.2 Current or Voltage Drive of the Laser

Up to now we assumed that a laser section was driven by a current I. The current density J in the carrier rate equation is then given by the injected current I divided by the active layer width w and length L_i of the electrode of that section. However, if the voltage V applied to the electrode of the section is given instead of I, then the current density is determined in a simplified way by

$$J(z) = [V - V_{DH}(N)]/(wL_t R_s) \quad (3.30)$$

$$V_{DH} = \frac{E_g}{q} + \frac{kT}{q}\left[2\ln\left(\frac{N}{N_S}\right) + C_1 N + C_2 N^2\right]$$

$$N_S = \sqrt{N_c N_v}$$

$$C_1 = \frac{\sqrt{2}}{4}\left(\frac{1}{N_c} + \frac{1}{N_v}\right) \quad C_2 = \left(\frac{3}{16} - \frac{\sqrt{3}}{9}\right)\left(\frac{1}{N_c^2} + \frac{1}{N_v^2}\right) \quad (3.31)$$

with N_c and N_v the effective densities of state of the conduction and valence bands, respectively, and E_g/q the bandgap potential [24]. $V_{DH}(N)$ represents the voltage drop across the double heterojunction and R_s is the electrical series resistance of the cladding layers of the section. In this case the current density J is not necessarily uniform across a section. Figure 3.7 illustrates the two basic situations for the laser drive.

In the remainder of this book we will continue to limit ourselves to current-driven lasers. However, this section clarifies that an extension toward voltage-driven laser sections is straightforward.

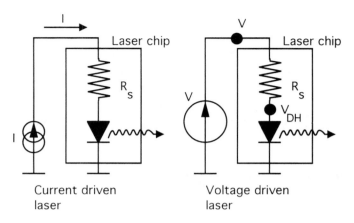

Figure 3.7 Schematic view of a current-driven and a voltage-driven laser.

3.7 THE STANDING-WAVE EFFECT IN GAIN-COUPLED LASERS

If a laser section contains a corrugated active layer as shown in Figure 3.8, then the modal gain Γg varies periodically. Gain coupling results here from the periodic variation of the confinement factor. Periodic variations of the carrier density and the local gain are washed out by the longitudinal carrier diffusion (with a diffusion length that is typically one order of magnitude larger than the corrugation period). Moreover, the gain coupling part of κ_{FB}, $\kappa_g = (\kappa_{FB} + \kappa_{BF}{}^*)/2$ with $\kappa_g = \kappa_g{}^*$, cannot be chosen freely. The gain coupling is indeed proportional to the gain that exists in the active layer. On the other hand, the threshold gain of the laser depends on the coupling coefficients. For a rectangular grating with duty cycle Λ_d/Λ, one can write

$$|\kappa_g| = \frac{1}{2\pi} \sin\left(\frac{\pi \Lambda_d}{\Lambda}\right) \frac{\Lambda}{\Lambda_d} (\Gamma_a - \Gamma_{min})g \qquad (3.32)$$

where Γ_a is the average optical confinement factor of the corrugated active layer, Γ_{min} the minimum confinement factor, and g the material gain in the active layer. For other grating shapes, formulas other than (3.32) need to be used [25].

If the modal gain varies periodically (due, for instance, to an active layer corrugation), then the stimulated emission term of the carrier rate equation (3.27) must be modified as explained here. For the case of gain gratings, the expansion of the local modal gain Γg in a Fourier series gives the following zero and first-order terms

$$\Gamma g = (\Gamma g)_a + 2\kappa_g \exp(-2j\beta_g z) + 2\kappa_g^* \exp(2j\beta_g z) \qquad (3.33)$$

with $(\Gamma g)_a$ the average modal gain and $\beta_B = 2\beta_g$ (first-order grating). For a first-order

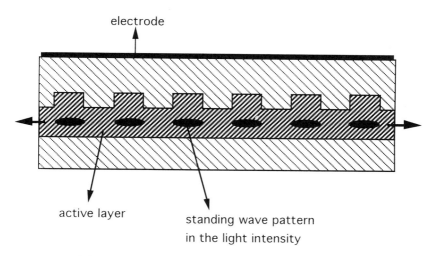

Figure 3.8 Gain-coupled DFB laser with a gain grating.

grating we can limit the expansion to the above two terms. Using the field expansion (3.1), the local optical field intensity can be written as

$$|E|^2 = \sum_m [|R_m^+|^2 + |R_m^-|^2 + R_m^+ R_m^{-*} \exp(-2j\beta_g z) + R_m^{+*} R_m^- \exp(2j\beta_g z)] \quad (3.34)$$

The net rate of stimulated recombination $\Gamma g|E|^2$ now becomes

$$\Gamma g \, |E|^2 = \sum_m [(\Gamma g)_a(|R_m^+|^2 + |R_m^-|^2) + 2\kappa_g R_m^{+*} R_m^- + 2\kappa_g^* R_m^+ R_m^{-*}] \quad (3.35)$$

As in (3.27), the spatially rapidly varying terms of $\Gamma g|E|^2$ have been neglected because the carrier diffusion length is large with respect to them. In gratings with an effective gain modulation, this modified stimulated emission must be used. It finds its origin in the overlap between the standing-wave pattern of the intensity of a mode and the gain grating. A mode that has a standing-wave pattern with intensity maxima at the points of high gain and minima at the points of low gain (see Figure 3.8) will receive on average a higher gain, but will on average also generate a higher stimulated recombination. The term $\kappa_g R_m^-(R_m^+)^* + (\kappa_g R_m^-)^* R_m^+$ expresses this overlap efficiency.

The overlap between the intensity of the standing-wave pattern and the gain grating differs from mode to mode and is responsible for the mode selectivity of gain gratings. This is called the *standing-wave effect*.

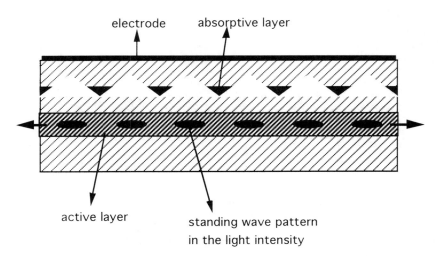

Figure 3.9 Gain-coupled DFB laser with a loss grating.

For loss gratings, the same mode selectivity mechanism applies. However, now a mode will on average suffer fewer losses if the mode has a standing-wave pattern with intensity maxima at the points of low loss and minima at the points of high loss. Figure 3.9 illustrates this. Because loss gratings are usually realized by a periodic thickness variation of an absorptive layer, the standing-wave effect has no impact on the stimulated recombination in the active layer, and the carrier rate equation remains valid in its original form (3.27).

3.8 BOUNDARY CONDITIONS

The coupled-wave equations and the carrier rate equation are valid only in a section with a uniform waveguide structure. In the case of lasers, consisting of several sections with different waveguiding properties, we can use the coupled-wave and carrier rate equations in each section separately. The fields and the carrier density in two neighboring sections A and B are then linked by the boundary conditions at the sections' interface (z_{AB}). The boundary conditions express the continuity of the tangential components of the electric and magnetic optical field vectors and of the carrier density.

In our model, no continuity of the carrier density along the laser axis can be imposed because no carrier diffusion is considered. However, the diffusion length in the longitudinal direction is sufficiently small (3 to 5 μm) to have large carrier density variations over relatively short axial distances in lasers of several hundred micrometers in length. Therefore, the boundary conditions can be ignored for N and discontinuities in the modeled carrier density are permitted.

The boundary conditions for the electromagnetic fields can be expressed in terms of reflection (r_j) and transmission (t_j) coefficients. For the fields at the boundary z_{AB} between two neighboring laser sections A and B, we get

$$\begin{pmatrix} E_m^+ \\ E_m^- \end{pmatrix}_{+z_{AB}} = \frac{1}{t_j} \begin{pmatrix} 1 & -r_j \\ -r_j & 1 \end{pmatrix} \begin{pmatrix} E_m^+ \\ E_m^- \end{pmatrix}_{-z_{AB}} \tag{3.36}$$

with $E_m^{\pm}$ defined by the modal components of (3.1), after integration over the (x, y)-plane and after applying the normalization integral (3.16). The reflection and transmission coefficients are determined by the difference in the effective refractive index between the neighboring sections. The sum of their squared moduli equals 1 if no radiative losses occur. They will be treated as given constants. For a discussion of the coupling efficiency between two axially coupled waveguides, the reader should refer to [26].

Equation (3.36) also applies to the inner and outer fields at the output facets, but if no light injection is present, the boundary conditions reduce to

$$E_m^+(z=0) = r_1 E_m^-(z=0); \quad E_m^-(z=L) = r_2 E_m^+(z=L) \tag{3.37}$$

with $r_{1,2}$ being the complex field reflection coefficient at the (left, right) facet.

In DFB laser theory, it is convenient to use complex field reflectivities. A random phase of the reflectivity can then account for the random variations in the phase of the grating at $z = 0$ and $z = L$. In this way a constant grating phase can be assumed at $z = 0$. In practice, this phase of the grating cannot be controlled with the current technological means and it varies from chip to chip. Figure 3.10 illustrates this phase

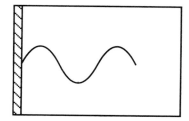
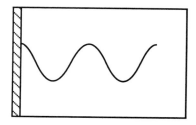

Figure 3.10 Uncertainty of the phase of the grating relative to a facet.

uncertainty of the grating versus the facets. As will be discussed in Chapter 6, this uncontrollable phase has an important impact on the DSM yield of DFB lasers.

References

[1] Baets, R., P. Kaczmarski, and P. Vankwikelberge, "Design and Modelling of Passive and Active Optical Waveguide Devices," NATO Advanced Study Institute on Waveguide Optoelectronics, Glasgow, 1990.
[2] Marcuse, D., *Theory of Dielectric Optical Waveguides*, New York: Academic Press, 1974.
[3] Agrawal, G., and N. Dutta, "Polarisation Characteristics of Distributed Feedback Semiconductor Lasers," *Appl. Phys. Lett.*, Vol. 46, February 1985, pp. 213–215.
[4] Agrawal, G., and N. Dutta, *Long Wavelength Semiconductor Lasers*, New York: Van Nostrand Reinhold, 1986.
[5] Morthier, G., and R. Baets, "Modelling of Distributed Feedback Lasers," Chapter 7 in *Compound Semiconductor Device Modelling*, C. M. Snowden and R. E. Miles, eds., New York: Springer-Verlag, 1992.
[6] Thompson, G., *Physics of Semiconductor Laser Devices*, New York: Wiley, 1980.
[7] Lax, M., "Fluctuations From the Nonequilibrium Steady State," *Rev. Mod. Phys.*, Vol. 32, January 1960, pp. 25–64.
[8] Lax, M., "Classical Noise IV: Langevin Methods," *Rev. Mod. Phys.*, Vol. 38, July, 1966, pp. 541–566.
[9] Henry, C., "Theory of Spontaneous Emission Noise in Open Resonators and Its Application to Lasers and Optical Amplifiers," *IEEE J. Lightw. Tech.*, Vol. 4, March 1986, pp. 288–297.
[10] Schubert, M., and B. Wilhelmi, *Nonlinear Optics and Quantum Electronics*, New York: Wiley, 1986.
[11] Jackson, J. D., *Classical Electrodynamics*, New York: Wiley, 1962.
[12] Yariv, A., *Quantum Electronics*, 3rd edition, New York: Wiley, 1989.
[13] Lee, D., *Electromagnetic Principles of Integrated Optics*, New York: Wiley, 1986.
[14] Vankwikelberge, P., J. P. Van de Capelle, R. Baets, B. H. Verbeek, and J. Opschoor, "Local Normal Mode Analysis of Index-Guided AlGaAs Lasers With Mode Filter," *IEEE J. Quant. Electron.*, Vol. QE-23, No. 6, June 1987, pp. 730–737.
[15] Yamamoto, Y., T. Kamiya, and H. Yanai, "Improved Coupled Mode Analysis of Corrugated Waveguides and Lasers," *IEEE J. Quant. Electron.*, Vol. QE-14, No. 4, April 1978, pp. 245–258.
[16] Streifer, W., D. R. Scifres, and R. D. Burnham, "Analysis of Grating-Coupled Radiation in GaAs:GaAlAs Lasers and Waveguides," *IEEE J. Quant. Electron.*, Vol. QE-12, No. 7, July 1976, pp. 422–428.
[17] Streifer, W., D. R. Scifres, and R. D. Burnham, "Coupled Wave Analysis of DFB and DBR Lasers," *IEEE J. Quant. Electron.*, Vol. QE-13, No. 4, April 1977, pp. 134–141.
[18] Kazarinov, R., and C. H. Henry, "Second-Order Distributed Feedback Lasers With Mode Selection Provided by First-Order Radiation Losses," *IEEE J. Quant. Electron.*, Vol. 21, 1985, pp. 144–150.
[19] Noll, R. J., and S. H. Macomber, "Analysis of Grating Surface Emitting Lasers," *IEEE J. Quant. Electron.*, Vol. 26, No. 3, March 1990, pp. 456–466.
[20] Casey, H., and M. Panish, *Heterostructure Lasers, Part A: Fundamental Principles*, New York: Academic Press, 1978.
[21] Petermann, K., "Calculated Spontaneous Emission Factor for Double-Heterostructure Injection Lasers With Gain-Induced Waveguiding," *IEEE J. Quant. Electron.*, Vol. 15, July 1979, pp. 566–570.
[22] Vankwikelberge, P., G. Morthier, and R. Baets, "CLADISS—A Longitudinal model for the Analysis

of the Static, Dynamic, and Stochastic Behavior of Diode Lasers With Distributed Feedback," *IEEE J. Quant. Electron.*, Vol. 26, No. 10, October 1990, pp. 1728–1741.

[23] Tucker, R., and D. Pope, "Circuit Modeling of the Effect of Diffusion Damping in a Narrow-Stripe Semiconductor Laser," *IEEE J. Quant. Electron.*, Vol. QE-19, No. 7, July 1983, pp. 1179–1183.

[24] Joyce, W. B., "Analytic Approximations for the Fermi Energy in (Al,Ga)As," *Appl. Phys. Lett.*, Vol. 32, No. 10, 15 May 1978, pp. 680–681.

[25] David, K., J. Buus, G. Morthier, and R. Baets, "Coupling Coefficients in Gain Coupled DFB Lasers: Inherent Compromise Between Coupling Strength and Loss," *IEEE Phot. Tech. Lett.*, Vol. 3, May 1991, pp. 439–441.

[26] Suematsu, Y., K. Kishino, S. Arai, and F. Koyama, "Dynamic Single-Mode Semiconductor Lasers With a Distributed Reflector," *Semiconductors and Semimetals*, Vol. 22, Lightwave Communications Technology, Pt. B, New York: Academic Press, 1985.

Chapter 4

Applying the Coupled-Mode Theory

The simplest application of the coupled-mode theory is the determination of the threshold of a laser. For simple DFB lasers and phase-shifted DFB lasers, the resonance condition, which determines the threshold state, can be described by an analytic expression, the so-called characteristic equation of the laser. This characteristic equation must in general be solved numerically, although analytical approximations exist for some special laser structures. The first section of this chapter is devoted to these analytical approximations and a discussion of a range of classic DFB laser structures: antireflection (AR)-coated DFB lasers with an index or gain grating, non-AR-coated DFB lasers, AR-coated $\lambda/4$ phase-shifted index-coupled DFB lasers, and second-order index-coupled DFB lasers.

Subsequently, different numerical solution techniques to solve the longitudinal coupled-mode model will be discussed in detail. Two general classes of solution techniques can be distinguished, one based on a narrowband approach and one on a broadband approach. In the narrowband approach, the fields are decomposed into longitudinal modes, each with a bandwidth below the inverse of the cavity round-trip time. The broadband solution technique consists of a large-signal dynamic analysis based on a simple spatial and temporal finite difference scheme for field amplitudes that includes the entire spectral bandwidth. This kind of approach is more suited for the study of short-pulse phenomena.

The above numerical methods take the longitudinal variations of the carrier density and the optical fields into account. Therefore, they go beyond the capabilities of the rate equation model of Chapter 2. This chapter also shows how the rate equations of Chapter 2 are linked to the longitudinal model of Chapter 3. In addition, the longitudinal spatial hole burning phenomenon is briefly explained.

Finally, this chapter shows how the coupling coefficients of a grating can be calculated. Coupling coefficients are essential parameters in the analysis of DFB lasers, since they describe the impact of the grating.

4.1 THRESHOLD SOLUTIONS FOR SIMPLE DFB LASERS

For a simple, single-section DFB laser, the threshold behavior can be derived semi-analytically from the coupled-mode equations. The laser threshold corresponds to the current level at which a transition from amplifier operation to oscillator operation occurs. This transition takes place when the population inversion in the active layer provides sufficient stimulated emission to compensate for the absorption and (distributed and discrete) feedback losses. At threshold, the presence of even the slightest photon density will deplete the carrier density and decrease the stimulated emission again below the losses. Therefore, the threshold situation corresponds to fulfillment of the resonance conditions without light propagation inside the cavity. In practice, the output power consists of amplified spontaneous emission at this current level.

In the theoretical threshold analysis, one can neglect the power, the Langevin forces, and any time dependencies. Furthermore, at threshold, a uniform carrier density can be considered in single-section DFB lasers due to a uniform carrier injection and the lack of any spatial hole burning. To this uniform carrier density corresponds a z-independent complex Bragg deviation $\Delta\beta_m$ for the mth longitudinal mode with wavelength λ_m. Given a Bragg deviation $\Delta\beta_m$, the static coupled-wave equations can be integrated analytically. Assuming a uniform grating (i.e., with a constant phase and amplitude for $\kappa_{FB,BF}$), the general relation between the fields at a point z and the fields at a point z_0 ($<z$) then becomes

$$\begin{pmatrix} R_m^+(z) \\ R_m^-(z) \end{pmatrix} = \begin{pmatrix} F_{11}^m(z-z_0) & F_{12}^m(z-z_0) \\ F_{21}^m(z-z_0) & F_{22}^m(z-z_0) \end{pmatrix} \begin{pmatrix} R_m^+(z_0) \\ R_m^-(z_0) \end{pmatrix}$$

$$F_{11}^m(z) = \cosh(\Delta_m z) - j\frac{\Delta\beta_m}{\Delta_m}\sinh(\Delta_m z)$$

$$F_{22}^m(z) = \cosh(\Delta_m z) + j\frac{\Delta\beta_m}{\Delta_m}\sinh(\Delta_m z)$$

$$F_{12}^m(z) = \frac{\kappa_{FB}}{\Delta_m}\sinh(\Delta_m z); \quad F_{21}^m(z) = -\frac{\kappa_{BF}}{\Delta_m}\sinh(\Delta_m z)$$

$$(\Delta_m)^2 = -[(\Delta\beta_m)^2 + \kappa_{FB}\kappa_{BF}] \tag{4.1}$$

The above matrix used for the solution of the coupled-wave equations is called a *propagator matrix* between z_0 and z [1,2]. In the next section we apply the above matrix equation to analyze the threshold behavior of different simple, single-section DFB laser structures.

4.1.1 AR-Coated Index- and Gain-Coupled DFB Lasers

The simplest threshold calculations are related to DFB lasers with both facets carrying AR coatings. In this case, the boundary conditions become very simple: $r_1 = r_2 = 0$. Setting $z = L$ and $z_0 = 0$ in (4.1) and applying the boundary conditions leads to the resonance condition for these lasers

$$\Delta_m \cosh(\Delta_m L) + j\Delta\beta_m \sinh(\Delta_m L) = 0 \qquad (4.2)$$

In general, one has to use numerical techniques to solve this equation for $\Delta\beta_m$.

The High-Gain Approximation

The high-gain approximation is valid for lasers with high feedback losses. In this case, the coupling coefficients are low and the threshold gain is high.
Equation (4.2) can also be rewritten as

$$\kappa_{FB}\kappa_{BF} \exp(2\Delta_m L) = (j\Delta\beta_m - \Delta_m)^2 \qquad (4.3)$$

In the limit where $\text{Im}(\Delta\beta_m) \gg \kappa$ with $\kappa = |\kappa_{FB}| = |\kappa_{BF}|$, the approximation $\Delta_m = -j\Delta\beta_m$ is valid. Equation (4.3) then reduces to

$$\kappa_{FB}\kappa_{BF} \exp(-2j\Delta\beta_m L) = -4\Delta\beta_m^2 \qquad (4.4)$$

A comparison of the phases of (4.4) gives us the phase resonance condition for $\text{Re}(\Delta\beta_m) \ll \text{Im}(\Delta\beta_m)$, that is, near the Bragg wavelength

$$\arg(\kappa_{FB}\kappa_{BF}) - 2\,\text{Re}(\Delta\beta_m L) = 2p\pi$$

$$p = 0, \pm 1, \pm 2, \ldots \qquad (4.5)$$

Index-Coupled DFB Lasers

For first-order index-coupled lasers, $\kappa_{FB} = -\kappa_{BF}^*$ and (4.5) then reduces to

$$\text{Re}(\Delta\beta_m L) = -(p + 1/2)\pi \qquad (4.6)$$

In this case, the two lowest order modes ($p = -1, 0$) have the resonance frequencies $\omega_B \pm \pi c/(2n_e L)$, where ω_B is the angular Bragg frequency: $\omega_B = 2\pi c/\lambda_B$. From (4.6), it is clear that for index-coupled DFB lasers, no resonance occurs at the Bragg wavelength. Moreover, (4.6) also indicates that the resonances are positioned symmetrically around the Bragg wavelength. Equating the amplitudes of (4.4) results in the amplitude resonance condition

$$\kappa^2 \exp[2 \, \text{Im}(\Delta\beta_m)L] = 4[\text{Re}(\Delta\beta_m)]^2 + 4[\text{Im}(\Delta\beta_m)]^2 \qquad (4.7)$$

This equation clearly indicates that the modes corresponding to $\text{Re}(\Delta\beta_m)$ values of the same magnitude but with opposite sign have the same gain ($\Gamma g - \alpha_{\text{int}} \sim \text{Im}(\Delta\beta_m)$). Therefore, first-order, index-coupled, AR-coated DFB lasers show multimode operation or degeneracy at their threshold.

Figure 4.1 illustrates the numerical solution of (4.2) in the ($\text{Re}(\Delta\beta_m)L$, $\text{Im}(\Delta\beta_m)L$) plane as a function of $\kappa = |\kappa_{FB}|$ for index-coupled lasers [3]. Comparison with the high-gain approximation shows that the latter is reasonably good for threshold gains $\exp[2 \, \text{Im}(\Delta\beta_m)L]$ in excess of 20 dB [4].

From Figure 4.2, a simple explanation for the absence of a mode at the Bragg wavelength λ_B ($\text{Re}(\Delta\beta_m) = 0$) can be derived. Indeed, consider the sum of distributed

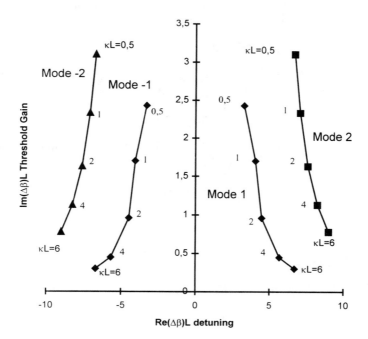

Figure 4.1 Solution of the threshold resonance condition for an AR-coated DFB laser in the ($\Delta\beta_r L$, $\Delta\beta_i L$)-plane. Different values for κL are considered. (*Source:* [3].)

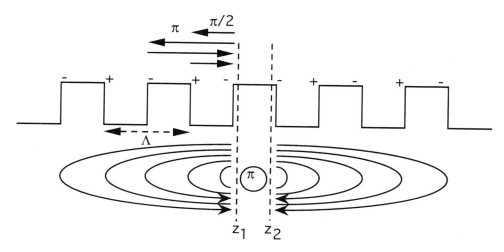

Figure 4.2 Physical explanation of the absence of laser resonance at the Bragg wavelength in an AR-coated DFB laser (*After:* [3]).

reflections in both directions at z_1 and z_2. At the Bragg wavelength, all the reflections at z_1 and z_2, add up constructively because the phases of subsequent interface reflections alternate between 0 and π and because the phase shift for propagating along half a period $\Lambda/2$ is $\pi/2$. The total round-trip phase for each wave will now be $(2p + 1)\pi$, with p being an integer. The additional phase shift π comes from the bidirectional propagation across $[z_1, z_2]$. As one observes, the round-trip phase is not a multiple of 2π at the Bragg wavelength, and therefore no mode occurs at this wavelength.

Gain-Coupled DFB Lasers

For first-order purely gain (loss)-coupled lasers, $\kappa_{FB} = \kappa_{BF}{}^*$ holds and (4.5) then reduces to

$$\mathrm{Re}(\Delta\beta_m)L = -p\pi \tag{4.8}$$

Equation (4.8) shows that this type of laser has a resonant mode ($p = 0$) at the Bragg wavelength and (4.7) indicates that this mode has the lowest threshold gain. The amplitude resonance condition (4.7) also shows us that the mode spectrum is symmetric with respect to the Bragg wavelength. Therefore, purely gain-coupled DFB lasers exhibit single-mode behavior. Physically, this is due to the standing-wave effect.

The presence of forward and backward propagating waves in the cavity gives rise to a standing-wave pattern in the optical power density with a period equal to half the wavelength inside the cavity (i.e., $\lambda/2n_e$). For the mode at the Bragg wavelength, this standing-wave pattern has a perfect overlap with the periodic gain variation (high (low) gain exists where the optical power is maximum (minimum)). A considerable increase of the average stimulated emission rate therefore results from the overlap of both periodic functions. For the other modes, however, the periods of the gain grating and of the standing-wave pattern in the optical power are different. It can be shown that there is no net effect on the average stimulated emission in this case.

The occurrence of a mode at the Bragg wavelength for purely gain (loss)-coupled lasers can be explained by means of Figure 4.2. For pure gain coupling, the phases of subsequent reflections alternate between $-\pi/2$ and $\pi/2$. At the Bragg wavelength, the total round-trip phase for each wave is now a multiple of 2π and a mode is therefore possible at the Bragg wavelength.

It must be remarked that the gain-coupling strength cannot be chosen freely for a laser with a corrugated active layer. The gain coupling is proportional to the gain that exists in the active layer in this case. However, the threshold gain itself depends on the coupling coefficients [5]. Furthermore, (4.7) (and also (4.3)) shows that the threshold gain usually decreases with increasing gain-coupling coefficient. Of course, the threshold gain also depends on the internal loss α_{int}. Figure 4.3 illustrates this situation. Curve a depicts the dependence of the gain-coupling coefficient on the average gain in the active layer and curve b depicts the dependence of the average threshold gain on the gain-coupling coefficient as it follows from the amplitude resonance condition. Curve a depends on the specific grating shape (e.g., depth, shape, and duty

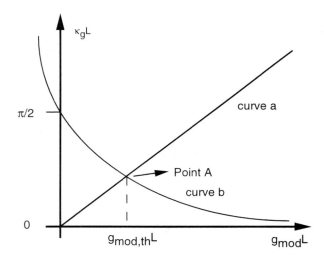

Figure 4.3 Relationship between $\kappa_g L$ and $g_{mod} L$ for a gain-coupled laser with a corrugated active layer.

cycle), but is linear in the modal gain for a given grating. Curve b, on the other hand, is independent of the grating shape, but depends on the internal losses (and also on the possible phase shifts, facet phases, and index coupling for the more complex DFB lasers considered in the subsequent sections).

If the gain coupling is realized with a loss grating, the above direct relationship between κ_g and the gain no longer exists. In other words, curve a in Figure 4.3 becomes horizontal. However, an extra average loss must now be added to the internal losses that already exist in the absence of the grating. This extra loss is given by

$$\alpha_{extra} = \frac{1}{\Lambda} \int_0^\Lambda \Delta\alpha(z)\, dz \qquad (4.9)$$

with $\Delta\alpha$ the local loss experienced by the waveguide mode ($\Delta\alpha = \Gamma_{grat}(z)\Delta\alpha_{mat}(z)$), which depends on the local confinement factor of the grating and on the variation of the material losses. For a rectangular grating with no material loss variations in the lossy layer, one finds

$$\alpha_{extra} = \frac{\Lambda_l}{\Lambda} \frac{2\pi}{\sin(\pi\Lambda_l/\Lambda)} \qquad (4.10)$$

with Λ_l/Λ being the duty cycle of the rectangular grating.

The minimization of the extra loss is of great importance to obtain acceptable threshold currents with loss gratings. The minimum value can only be obtained with rectangular or triangular gratings with a very small duty cycle. A grating with, for instance, a sinusoidal form already gives an extra loss of $4|\kappa_g|$.

In general, gain-coupled lasers only show a partial gain coupling. In this case the mode spectrum is no longer symmetric, but the gain coupling still ensures nondegeneracy. In the remainder of this book the focus will be on index-coupled lasers, although most of the results can be extended to gain-coupled lasers.

Low-Gain Approximation

The low-gain approximation is applicable when $\text{Im}(\Delta\beta_m) \ll \kappa$. Equation (4.3) then reduces to [6]

$$\text{Re}(\Delta\beta_m)L \approx \pm\kappa L, \quad \text{Im}(\Delta\beta_m)L \approx (\pi/\kappa L)^2 \qquad (4.11)$$

for the threshold modes of index-coupled DFB lasers. Comparison with numerical data indicates that this is a good approximation for threshold gains below 5 dB [4]. For purely gain-coupled lasers the low-gain approximation gives

$$\text{Re}(\Delta\beta_m)L = 0, \quad \text{Im}(\Delta\beta_m)L = -\kappa_g L + \frac{\pi^2}{2}\frac{\kappa_g L}{(1+\kappa_g L)^2} \quad (4.12)$$

for the threshold mode.

4.1.2 The Stopband or Energy Gap

Wave propagation in periodic real refractive index structures gives rise to a stopband or energy gap. This is a phenomenon well known in the study of crystals. The periodic grating causes a gap in the spectrum of propagating waves. Figure 4.4 illustrates the wave vector versus frequency plot in a lossless waveguide with an infinitely long index grating. The width of the stopband or energy gap expressed in terms of $\text{Re}(\Delta\beta)$ is equal to 2κ. Wave propagation only occurs outside the gap. Inside the gap or stopband, the waves exponentially decay and hence can propagate only over a finite distance. Therefore, in gratings of finite length, the decaying waves tend to occur only around places where the periodic grating is interrupted, such as at a boundary. The decay is due to the distributed reflections occurring when the coupling becomes strong. Waves outside the gap are traveling waves that show little interaction with the periodic structure.

By introducing gain in the waveguide, the decay can be overcome. However, in the stopband of an AR-coated index-coupled laser, no mode appears because no resonance is possible, as illustrated by Figure 4.2. As the next sections will show, a mode in the energy gap is possible after introduction of facet reflectivities, phase shifts, or gain or loss gratings. Such lasers with a mode in the energy gap tend to have high mode selectivity for relatively short structures ($\kappa L \approx 1$).

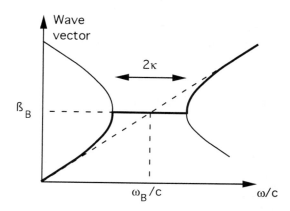

Figure 4.4 Dispersion diagram for an infinite index grating in a waveguide without loss or gain.

In gain-coupled gratings, no frequency stopband occurs. On the other hand, there is now a forbidden band of width 2κ in the wave vector. This can be represented by interchanging the axes of Figure 4.4.

4.1.3 DFB Lasers With Reflecting Facets

The presence of reflecting facets will very often remove the mode degeneracy and symmetry encountered in the previous section for index-coupled lasers. When the restriction of AR coatings is dropped and the boundary conditions for the fields $E^\pm$ are taken into account, the laser resonance condition becomes

$$[r_2 r_1 F^m_{11}(L) + r_2 F^m_{21}(L)] \exp(-2j\beta_g L) = r_1 F^m_{12}(L) + F^m_{22}(L) \qquad (4.13)$$

In general, one has to use numerical techniques to solve this equation for $\Delta\beta_m$. As explained in Chapter 3, the field reflection coefficients r_1 and r_2 show random phases. Figure 4.5 shows the solutions of (4.13) for a DFB laser with one reflecting facet

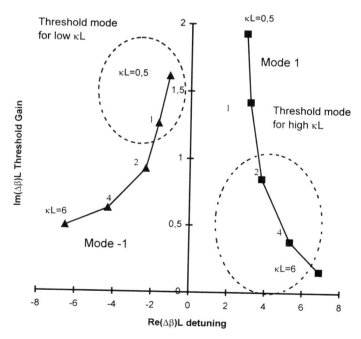

Figure 4.5 Solutions of the threshold resonance condition for a DFB laser with one AR-coated facet ($R_2 = 0$) and one uncoated facet ($R_1 = 0.32$, phase 0), in the ($\Delta\beta_r L$, $\Delta\beta_i L$) plane. Different values for κL are considered. (*Source:* [3].)

($|r_1|^2 = R_1 = 0.32$, phase 0). The spectrum is no longer symmetric and is also no longer degenerate.

Whether an index-coupled DFB laser with reflecting facets will operate in a single mode strongly depends on the amplitude of the reflection coefficients and on their random phases. The latter have a strong impact on the difference in threshold gain between the lowest order longitudinal mode and the nearest side mode [7,8]. This threshold gain difference is a measure of single-mode behavior. The mode selectivity arises here from the interference of the discrete facet reflections and the distributed reflections from the grating. Maximum interference and hence maximum selectivity occurs when the two contributions are of comparable strength. Therefore, larger κL values (≈ 3) are required in lasers with both facets cleaved than in lasers with one facet cleaved and one coated mirror. In the latter case, a lower distributed feedback and κL value (≈ 1) are more suitable for interfering with the lower facet feedback.

Additional mode selectivity can be introduced by adding some gain coupling to the DFB laser with facet reflections.

4.1.4 The λ/4 Phase-Shifted DFB Laser

We already identified two methods to avoid mode degeneracy: use gain coupling or introduce facet reflections. Alternatively, the mode degeneracy in AR-coated index-coupled DFB lasers can also be removed by creating a λ/4 phase shift in the grating. This approach is more predictable than the use of reflecting facets, but the fabrication of phase-shifted lasers is more complex.

As shown in Figure 4.6, the introduction of a λ/4 phase shift corresponds with

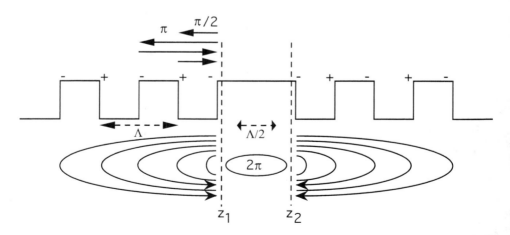

Figure 4.6 Physical explanation of the phase resonance at the Bragg wavelength in an AR-coated λ/4 phase-shifted DFB laser. (*After:* [3].)

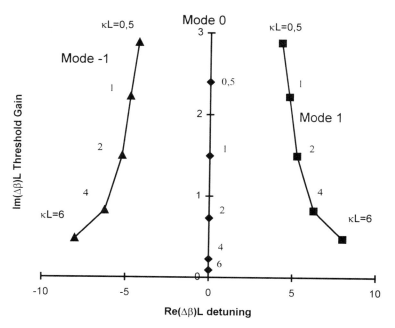

Figure 4.7 Solutions of the threshold resonance condition for an AR-coated $\lambda/4$ phase-shifted DFB laser in the ($\Delta\beta_r L$, $\Delta\beta_i L$) plane. Different values for κL are considered. (*Source:* [3].)

the insertion of a phase shift section of length $\Lambda/2 = \lambda/(4n_e)$ in the center of the grating structure. This phase shift region creates an extra $\pi/2$ phase shift (between $[z_1, z_2]$ in Figure 4.6) for each wave passing along it. It is clear that the round-trip phase shift will now receive an extra phase shift of π compared with the situation of Figure 4.2.

The resonance condition is derived by calculating the reflection coefficient r_i at the inner end of a laser section of length $L/2$ that is AR-coated at the outer end. Assuming that the laser consists of two such sections separated by a phase shift of $\pi/2$ then gives the round-trip resonance condition

$$\left[\Delta_m \coth\left(\Delta_m \frac{L}{2}\right) + j\Delta\beta_m\right]^2 = -\kappa_{FB}\kappa_{BF} \qquad (4.14)$$

Figure 4.7 illustrates the solutions of (4.14) in the $(\text{Re}(\Delta\beta_m)L, \text{Im}(\Delta\beta_m)L)$ plane as a function of $\kappa = |\kappa_{FB}|$ [3]. One can clearly observe the single lowest order mode at the Bragg wavelength.

The single-mode behavior at threshold may get lost once far above threshold due to longitudinal spatial hole burning [9]. Therefore, multiple phase-shifted DFB lasers have also been designed. Their main objective is to reduce the amount of spatial hole burning by rendering the total power distribution of the lasing mode more uniform along the laser while tolerating higher values for the κL product. A thorough treatment of $2 \times \lambda/8$ phase-shifted AR-coated lasers is given in [10].

4.1.5 Second-Order Index-Coupled DFB Lasers

From Section 3.5.2, we learn that the static coupled-wave equations (without any noise terms) for a second-order index-coupled DFB laser can be written as [11,12]

$$\frac{dR_m^+}{dz} + (j\Delta\beta_m + \kappa_r) R_m^+ = (-\kappa_r + j\kappa_i)R_m^-$$

$$-\frac{dR_m^-}{dz} + (j\Delta\beta_m + \kappa_r) R_m^- = (-\kappa_r^* + j\kappa_i^*)R_m^+ \qquad (4.15)$$

with $j\kappa_i$ a measure of the strength and phase of the second-order index-coupling. The κ_r terms in (4.15) correspond to a first-order excitation of radiation modes whose influence can be described by a periodically varying absorption. The period of this absorption loss is half the period of the second-order grating. Therefore, as (4.15) indicates, a second-order index-coupled DFB laser behaves similarly to a partly gain-coupled first-order DFB laser with a loss grating. However, while the value of κ_r is typically limited to 5 to 10 cm^{-1}, loss gratings allow gain-coupling coefficients as high as 50 cm^{-1}.

Another main difference between second-order DFB lasers and loss-coupled lasers resides in the extra losses. In Section 4.1.1 we saw that a loss grating causes an extra average loss α_{extra}. This extra loss is proportional to the loss coupling coefficient κ_g, but the ratio between both depends on the specific form of the grating. On the other hand, (4.15) shows that for second-order index-coupled lasers an extra power loss 2 Re(κ_r) appears. This loss represents the extra average loss due to the radiation modes. The ratio of this extra loss to κ_r is independent of the shape of the second-order grating.

As in partly gain-coupled lasers, the periodic loss variation causes additional mode selectivity [11]. This can be explained as follows. From (4.15) the following power relation can be derived for a longitudinal mode m [12]:

$$P_{out,m} = 2 \operatorname{Im}(\Delta\beta_m) \int_0^L (|R_m^+|^2 + |R_m^-|^2)\,dz - 2\int_0^L \operatorname{Re}[(R_m^- + R_m^+)^*(\kappa_r R_m^- + \kappa_r^* R_m^+)]\,dz$$

$$(4.16)$$

The left-hand side of (4.16) represents the output power of the laser at both facets. The first term of the right-hand side corresponds to the power generated by stimulated emission minus the internal absorption (caused by the unperturbed waveguide losses), while the second term corresponds to the power radiated by the superposition of the first-order (90°) scatterings of the forward and backward waves of the mode considered. The magnitude of the radiated power will depend on the interference between the contributions of the forward and backward waves. If the contributions interfere destructively, the radiated power will be low and the mode will show little radiation loss. However, when constructive interference appears, strong radiation losses will appear.

Using the low-gain approximation ($|\kappa_i|L \gg \pi$) and assuming small $|\kappa_r|$ ($\ll |\kappa_i|$), [12] shows that for AR-coated second-order DFB lasers, the mode with Bragg deviation $\mathrm{Re}(\Delta\beta_m) = \kappa = |{-\kappa_r + j\kappa_i}|$ has the lowest threshold gain. For this mode, destructive interference occurs. The mode with the second lowest threshold gain appears on the other (long wavelength) side of the stopband (with $\mathrm{Re}(\Delta\beta_m) = -\kappa$). This mode suffers from constructive interference. Moreover, under the above assumptions, the threshold gain difference can now be approximated by

$$\Delta g L = 4\,\mathrm{Re}(\kappa_r)L \qquad (4.17)$$

4.2 NUMERICAL SOLUTIONS OF THE COUPLED-MODE MODEL

The full longitudinal model described in the previous chapter consists of the coupled-mode equations, the carrier rate equations, and the boundary conditions. In general, the solutions of this set of highly nonlinear partial differential equations can only be determined through numerical techniques.

Two important approaches exist for solving the coupled-mode model: a narrowband approach and a broadband approach. In the coupled-mode model of Chapter 3, each longitudinal mode is represented by its own carrier wave with carrier (reference) frequency ω_m. Consequently, (3.1) expresses the optical field as a set of different carriers. The phasor $R_m^{\pm}(z, t)$ associated with each mode is still time-dependent, but its temporal variations due to noise, instabilities, or modulation are slow compared to the optical frequency. The dynamic or static spectral width of each phasor corresponds to the spectral width of the associated mode. By using one phasor per mode, the representation of the full spectral width of the laser is compressed in a narrowband that corresponds in size to the broadest mode. For high-speed modulations, a total bandwidth of a few gigahertz up to a few tens of gigahertz is then required to model the laser. This compression in the spectral representation does not come for free. Spectral width has been traded for an increased number of phasors or field variables. In this way the coupled-mode description of Chapter 3 is a narrowband approach.

Another way to analyze lasers is to consider only a single carrier wave that carries the full spectral width of the laser. In this case, (3.1) only contains one term in the summation of phasors. The single phasor then represents all the modes that are taken into account. Such a representation requires the spectral bandwidth covered by the model to equal the full spectral width of the laser. This approach is called a *broadband approach*. If p modes are considered, the total modeled spectral bandwidth must be on the order of $(p-1)\delta\omega_m/2\pi + \Delta\nu_{dyn}$, where $\delta\omega_m/2\pi$ is the mode spacing of the laser and $\Delta\nu_{dyn}$ is the dynamic spectral width of the broadest mode. For a laser of length 300 µm and an effective index of 3.25, the mode spacing is on the order of 150 GHz. For an analysis considering only three modes, this already requires a spectral bandwidth of about 310 to 350 GHz.

Although the model of Chapter 3 was derived for a multimode/multiphasor field representation, the model stays basically valid for a single-phasor representation. The slowly varying amplitude approximation of Chapter 3 remains valid because the optical frequencies are on the order of 200 THz, which is still considerably higher then the full spectral width of the laser. Moreover, the spatial variations for the multiple- or single-phasor approach are similar in size. A more complicated issue is the wavelength dependence of the gain, the gain suppression due to spectral hole burning, and the Bragg deviation $\Delta\beta_{m,r}$. These parameters tend to vary considerably within the spectral window of interest. In the narrowband approach, the different reference wavelengths λ_m corresponding to the angular reference frequencies ω_m ($\lambda_m = 2\pi c/\omega_m$) are used in the wavelength-dependent coefficients of the modal wave equations and the carrier density equation. As will be discussed further on, in the broadband approach the instantaneous wavelength is used.

Since a narrowband model only covers a baseband frequency range f_B of a few gigahertz up to a few tens of gigahertz , calculating the dynamic behavior requires a sampling period $(2f_B)^{-1}$ of a few tens of picoseconds. Broadband models, on the other hand, may cover a baseband frequency range f_B of up to 1,000 GHz. In that case a calculation of the laser dynamics needs a sampling period as small as 0.5 ps. Note that the sampling periods mentioned here are the maxima implied by Nyquist's theorem. A numerical algorithm may require smaller periods for reasons of numerical stability or accuracy.

4.3 THE NARROWBAND APPROACH FOR SOLVING THE COUPLED-MODE MODEL

The multimode model derived in Chapter 3 already implies a narrowband approach. This is also the most frequent type of model used for laser diodes. It does, however, require that one can assess ahead of time how many modes must be taken into account when applying the model to a certain laser. Considering too many modes makes the calculations unnecessarily heavy; using not enough modes leads to incorrect results.

In the narrowband approach, the coupled-mode model can be solved in different ways depending on the type of analysis that is desired. Each type of analysis is designed to calculate in an optimized way certain laser properties under specific drive conditions. Therefore, it is important to question the kind of laser properties and the kind of external drivers that one is interested in. Table 4.1 groups laser characteristics and external drive forms according to different types of analysis. By focusing on a limited combination of characteristics and external drivers per type of analysis, a simplification of the numerical algorithms required can often be obtained. In this way, each type of analysis uses numerical techniques that are specific to it and that introduce some constraints.

In the remainder of this section, the different types of analysis mentioned in Table 4.1 are discussed. A more detailed explanation of the solution methods can be found in [13]. As will also become clear, solutions obtained with one type of analysis may often be useful to other types. The techniques discussed here have been implemented in the laser device simulator CLADISS (compound cavity laser diode simulation software) [14,15]. All the calculated data in this book based on the longitudinal coupled-mode model were obtained using the CLADISS computer model.

Note that a computer model is referred to as longitudinal if the detailed longitudinal spatial variations of the carrier density and of the optical fields are taken into account. Other longitudinal narrowband computer simulations, based on the model of Chapter 3, have been presented in the literature [10,16–18].

Table 4.1
Different Types of Analysis Within the Narrowband Approach

Type of Analysis	Constraints	Calculated laser characteristics	Drive properties
Threshold analysis	•Static •No noise sources •No optical power	•Threshold current •Threshold wavelength •Threshold gain	Static current injection
Static continuous wave analysis	•Static •No noise sources	•P/I curve •λ/I curve	Static current injection
Small-signal dynamic analysis	•No noise sources •Small-signal dynamics	•IM response •FM response •Second-, third-order distortions in FM and IM response	Static bias current with sinusoidal modulation on top
Noise analysis	•Small-signal dynamics	•RIN •Linewidth	Static current injection
Large-signal dynamic analysis	•Noise sources optional •No subpicosecond behavior	•Step response •Impulse response •Power spectrum optional	Large-signal injection current variations

4.3.1 Threshold Analysis

The threshold analysis serves to investigate at which current level a transition from amplifier operation to oscillator operation of the laser occurs.

In a threshold analysis, the time dependencies, the stimulated recombination rate in the carrier rate equation, the power dependence of the gain, the Langevin forces, and the spontaneous emission in the coupled-wave equations are all neglected. The carrier rate equation (3.27) for a laser section with uniform waveguide geometry and uniform current injection is then reduced to a third-order polynomial equation in N. The coupled-wave equations (3.11), in which the Langevin forces and the time dependencies are ignored, can be solved for this uniform carrier density. For a uniform grating (i.e., with a constant phase and amplitude for $\kappa_{FB,BF}$), this leads to equations of the type in (4.1).

Equations (4.1), together with the reduced form of the carrier rate equation (3.27) and the boundary conditions (3.36) and (3.37), can now be used to calculate the laser threshold current and wavelength. One way to do this is given below.

Consider a general multisection laser as shown in Figure 4.8. The current densities injected into the different sections are assumed to depend linearly on one independent current density J_v injected into section v. The threshold analysis searches for those values of J_v for which at least one modal frequency ϖ_m ($= \omega_m + \Delta\omega_m$) exists such that the amplitude and phase resonance conditions are fulfilled.

The amplitude and phase resonance conditions can be defined as follows. Take some point z_v somewhere along the cavity. As Figure 4.8 shows, the cavity is then divided into two parts, which can be replaced by effective reflection coefficients ρ_L for the left-hand part and ρ_R for the right-hand part. Those reflection coefficients depend

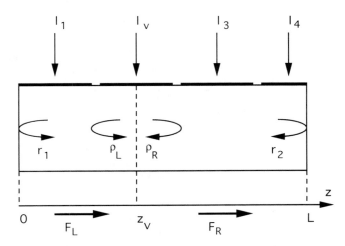

Figure 4.8 Division of a multisection DFB laser in two parts for the calculation of the round-trip gain.

on the frequency ϖ_m and on the current density J_v. Using the complex round-trip gain $\rho_L\rho_R$, the phase and amplitude resonance conditions can be expressed by

$$\rho_L(\varpi_m, J_v)\, \rho_R(\varpi_m, J_v) - 1 = 0 \quad (4.18)$$

The threshold analysis is now reduced to the search of the roots $(\varpi_{m,th}, J_{th,m})$ of (4.18). The couple with the lowest threshold current density $J_{th,m}$ defines the laser threshold and the main lasing mode. An illustration of the complex round-trip gain as a function of wavelength is shown in Chapter 6. The reflection coefficients ρ_L and ρ_R are calculated by means of the propagator matrices $F_L(0 \to z_r)$ and $F_R(z_r \to L)$ of the left-hand and right-hand parts of the cavity (Figure 4.8)

$$\rho_L = \frac{r_1(F_L)_{11} + (F_L)_{12}}{r_1(F_L)_{21} + (F_L)_{22}}\, e^{-2j\beta_g z_r}$$

$$\rho_R = \frac{(F_R)_{21} - r_2(F_R)_{11}e^{-2j\beta_g L}}{(F_R)_{22} - r_2(F_R)_{12}e^{-2j\beta_g L}}\, e^{2j\beta_g z_r} \quad (4.19)$$

Because the stimulated recombination is omitted from the carrier rate equation in the threshold analysis, the propagator matrices can be calculated straightforwardly without any iteration techniques once a frequency ϖ_m and a set of section current densities are specified.

Equation (4.18) can be solved in different ways. One useful method is the Newton-Raphson (NR) iteration procedure. The initial estimates needed to start this procedure can be obtained by looking for phase resonances with an amplitude close to 1 in a sufficiently wide frequency window in the (ϖ_m, J_v)-plane. To this end, the current is gradually changed from low values upward using a bisection approach. The derivatives $\partial\rho_{L,R}/\partial\varpi_m$ and $\partial\rho_{L,R}/\partial J_v$ required for the NR technique can be calculated analytically using (4.1) and (4.18).

4.3.2 Continuous-Wave Analysis

The CW analysis is a static, above-threshold analysis in which one can no longer neglect the stimulated recombination rate in the carrier rate equation or the power dependence of the gain. Since the optical power is generally not uniform in the longitudinal direction, a nonuniform carrier density (and hence a nonuniform Bragg deviation) will most probably occur.

The CW analysis aims at calculating the total output power and the central wavelength of each line in the optical spectrum as a function of static drive currents. It also delivers the internal field and carrier distributions above threshold. In the CW analysis, the laser is assumed to have reached a static regime (time-independent in the slowly varying time approximation of Chapter 3) and all derivatives with respect to time are thus neglected. Langevin forces, which determine the fluctuations in the output power and the form of each spectral line, are not included either, except for the average spontaneous emission that couples into the modes. Therefore, the field equations (3.26) can be used with the time derivatives removed and the notation $A_m^\pm$ replaced by $R_m^\pm$. The behavior of the carriers is given by (3.27) with the time derivatives dropped.

Given the static current excitations, the frequencies ϖ_m and power levels of the laser modes can be found by transforming the time-independent boundary value problem, (3.26), (3.27), (3.36), and (3.37), into an initial value problem that is solved with an iterative shooting method. This shooting method is based on the propagator matrix formalism.

To take into account the spatial variations of the carrier density along the cavity, each section of the laser can be divided into many small segments in which the carrier density is assumed uniform. A segment is typically a few microns long. For the ith segment $[z_i, z_{i+1}]$, the carrier density is then approximated by the value N_i in the middle z_i' of the segment. The N_i-value is found by solving the carrier rate equation at z_i'. The stimulated emission term in the carrier rate equation also contains the modal powers at z_i'. These modal powers are approximated by the average $P_m^\pm(z_i')$ of the power levels at z_i and z_{i+1}. For known drive currents and optical power levels, the discretized carrier rate equation again reduces to a third-order polynomial equation.

The average optical power $P_m^\pm(z_i')$ is also used to approximate the spontaneous emission in the propagator matrices. Due to the spontaneous emission, the propagator matrix description of (4.1) needs to be slightly adapted. For a segment of length $l_i = z_{i+1} - z_i$, we get

$$\begin{pmatrix} R_m^+(z_{i+1}) \\ R_m^-(z_{i+1}) \end{pmatrix} = s_c(l_i) \begin{pmatrix} F_{11}^m(l_i) & F_{12}^m(l_i) \\ F_{21}^m(l_i) & F_{22}^m(l_i) \end{pmatrix} \begin{pmatrix} R_m^+(z_i) \\ R_m^-(z_i) \end{pmatrix}$$

$$s_c(l_i) = \exp\left\{ \frac{R_{sp,m} h\nu_m l_i}{8L} \left(\frac{1}{P_m^+(z_i')} - \frac{1}{P_m^-(z_i')} \right) \right\} \quad (4.20)$$

Moreover, the coefficients F_{ij}^m of (4.20) correspond to the coefficients F_{ij}^m of (4.1) but with $j\Delta\beta_m$ replaced by

$$j\Delta\beta_m - \frac{R_{sp,m} h\nu_m}{8L} \left(\frac{1}{P_m^+(z_i')} + \frac{1}{P_m^-(z_i')} \right) \quad (4.21)$$

The shooting method now proceeds as follows. First, for each mode an initial estimate is introduced for its lasing frequency ϖ_m and its backward field $R_m^-(0)$, which can be taken as real without any restriction. Applying the left boundary condition gives the starting values for $R_m^+(0)$. The backward and forward waves of all the modes are then propagated through the laser taking the photon carrier interaction into account self-consistently. Therefore, the propagation through each segment $[z_i, z_{i+1}]$ requires a local iteration. Once arrived at the right facet ($z = L$), the second boundary condition requires $R_m^-(L) - r_2 R_m^+(L) \exp(-2j\beta_g L)$ to be zero. Starting from the initial estimates, an NR technique can be applied to find a solution for $(\varpi_m, R_m^-(0))$.

4.3.3 Small-Signal Dynamic Analysis

The usual approach toward a small-signal analysis is the sinusoidal regime. Such an analysis gives the small-signal IM and FM responses as functions of the modulation frequency $\Omega/2\pi$ (further on also referred to by the angular modulation frequency Ω). If the small-signal analysis includes higher order harmonic components, harmonic distortion can also be calculated. Note that in the small-signal analysis, noise sources are ignored, except for the average spontaneous emission.

The sinusoidal regime implies a phasor notation. However, before complex phasors related to the modulation frequency $\Omega/2\pi$ can be introduced, the complex number notation with respect to the optical reference frequencies ω_m must be removed. Therefore, the complex phasors $R_m^\pm$ must be replaced by their moduli $r_m^\pm$ and phases $\varphi_m^\pm$. This also implies that the wave equations for $r_m^\pm$ and $\varphi_m^\pm$ must be used, which corresponds to converting (3.26) into the (3.21)–(3.22) format.

In the sinusoidal regime, the injected current densities are expressed as

$$J_v(t) = J_{v0} + J_{v1} \cos(\Omega t) \tag{4.22}$$

with J_{v0} being the static bias current density and J_{v1} the sinusoidal modulation current density injected into section v. The periodic excitation makes the assumption of a periodic response plausible, and therefore the field quantities and the carrier densities can be expressed as a complex Fourier series with base pulsation Ω. The Fourier expansion can be limited to the first three terms, which is acceptable for sufficiently small modulation depths. Indeed, experiments [19] and theory [20] show that the power of the kth harmonics is proportional to the kth power of the modulation depth.

The laser equations are solved by substitution of the expansions, whereby the $\exp(jk\Omega t)$ time variation removes the time derivatives. Multiplication of the resulting equations with $\exp(-jk\Omega t)$ and integration over the time period corresponding to Ω leads to separate equations for each different index k. These separate equations for each index k can be written as a set of first-order differential equations in z. We will not give these equations because they are too extended and their derivation is straightforward. By applying finite differences with the same longitudinal discretization scheme as in the static analysis, these equations can be transformed into a set of linear algebraic equations. The discretized equations, together with the boundary equations, can be solved by standard techniques [13].

For a first-order small-signal analysis, k is limited to $k = 1$ and the nonhomogeneous source terms depend on the modulation current. For a second-order analysis, two sets of linear algebraic equations can be distinguished, one for the first-order harmonics ($k = 1$) and one for the second-order harmonics ($k = 2$). The set of equations corresponding to $k = 1$ is the same as the set of equations for the first-order small-signal analysis and it can be solved immediately. The set corresponding to $k = 2$ has first harmonics in its source terms. Therefore, it can only be solved after the first harmonics have been calculated. Similarly, for the third-order small-signal analysis, the third-order harmonics can only be calculated after the second-order harmonics have been determined.

It must be noted that such a solution may not represent the only possible type of solution of the dynamic laser equations for sinusoidal excitations. Indeed, due to the nonlinearity of these equations, a periodic excitation does not necessarily imply a periodic response with the same period. In fact, responses with different periods (e.g., period doubling, patterning effects) and nonperiodic responses have already been observed experimentally [21]. Such effects nevertheless tend to appear for large modulation currents and at high modulation frequencies. We therefore restrict ourselves to relatively small modulation depths, for which the expansion can be used.

4.3.4 Noise Analysis

The noise analysis is restricted here to situations with static injection only. The fluctuations, mainly caused by spontaneous emission, are nevertheless time-dependent, and therefore the time-dependent phase equations (3.22), intensity equations (3.24), and carrier equation (3.27) must be used. The Langevin forces, which represent the noise sources, can thereby be regarded as time-dependent excitations with a small amplitude.

The small amplitude of these Langevin forces will in general result in small fluctuations of the field amplitudes and phases and of the carrier density. Hence, a linearization of the laser equations into these small fluctuations is justified. In order to eliminate the derivatives with respect to the time, the Fourier transforms of the fluctuations of the field variables can now be introduced.

Since the Langevin functions can be considered as stationary for time differences on the order of 0.01 ns or more, their Fourier transforms at different Fourier frequencies are uncorrelated. In general, as long as only the spectral components in a bandwidth on the order of a few tens of gigahertz are considered, the correlation of the Fourier transforms $F_1(\Omega)$ and $F_2(\Omega)$ of two Langevin functions $f_1(t)$ and $f_2(t)$, as found in laser diodes, is given by

$$\langle F_1(\Omega) F_2^*(\Omega') \rangle = 2D_{12} \delta(\Omega - \Omega') \qquad (4.23)$$

Applying the Fourier transform to the linearized equations results again in a set of differential equations in z for the Fourier transforms of the fluctuations of the field and carrier variables. The obtained system of differential equations is identical to that obtained in the first-order small-signal analysis, except that now other variables occur and other nonhomogeneous terms (i.e., the Fourier transforms of the Langevin functions instead of the modulation currents) are present as source terms [13].

By discretizing the field variables and the carrier density, this system of differential equations can be reduced to a linear algebraic system similar to the one obtained in the small-signal dynamic analysis. For each pair of discrete (field or carrier) quantities $X_1(\Omega)$ and $X_2(\Omega)$, the second-order moment $\langle X_1(\Omega) X_2^*(\Omega') \rangle$ can now readily be determined using (4.23) and the Gaussian character of the Langevin forces (see Section 3.4.3). These second-order moments can then be further transformed into the FM and intensity noise spectra. The power spectrum and the linewidth can then be derived from the spectral density of the FM noise as explained in Chapter 2 and [22].

4.3.5 Large-Signal Dynamic Analysis

The equations used for large-signal analysis are the time-dependent coupled-wave equations (3.26) and the time-dependent carrier rate equation (3.27). These time-dependent equations are discretized for the time variable t using the approximations $\partial R_m^\pm / \partial t = (R_m^\pm(z, t) - R_m^\pm(z, t - \Delta t))/\Delta t$ and $\partial N/\partial t = (N(z, t) - N(z, t - \Delta t))/\Delta t$. The resulting equations can then be transformed into equations similar to the those for the static analysis. These equations are solved for each instant in time (the different moments being separated by a time step Δt) in a similar way to that done in the static analysis. This approach assumes that the considered dynamic variations are slow with respect to the cavity round-trip time. From the fields at $z = 0$ and $z = L$ and the frequency, the output powers and the chirp can be obtained as a function of time. Chirp corresponds to dynamic line broadening due to frequency modulation. Langevin functions can be included in the analysis by means of randomly generated numbers with the appropriate statistics. The spectral width can then be calculated using the fast Fourier transform (FFT).

4.4 THE BROADBAND APPROACH FOR SOLVING THE COUPLED-MODE MODEL

Some authors have used a broadband approach for solving the coupled-mode equations [23,24]. As already mentioned in Section 4.2, the coupled-mode model of Chapter 3 remains valid. However, the instantaneous wavelength is used in the gain function:

$$\lambda(t) = \lambda_{\text{ref}} - (\lambda_{\text{ref}}^2/2\pi c)\, \partial \phi^-(0, t)/\partial t \qquad (4.24)$$

with λ_{ref} chosen close to the expected lasing wavelength.

In the broadband approach the laser equations are solved in the time domain using a large-signal transient analysis. To this end, the partial differentials in the wave equations are discretized using a first-order finite difference schema. This gives

$$R^+(z + \Delta z, t + \Delta t) = [1 - j\Delta\beta\Delta z]R^+(z, t) + j\kappa_{FB}\Delta z R^-(z, t) + F^+$$

$$R^-(z, t + \Delta t) = [1 - j\Delta\beta\Delta z]R^-(z + \Delta z, t) + j\kappa_{BF}\Delta z R^+(z, t) + F^- \quad (4.25)$$

The spectral behavior can be observed by using FFT to convert the time-domain signals in the optical frequency domain.

The broadband approach has some relevant advantages. Its calculations based on (4.25) are simple and fast and it shows no convergence problems. It is also better suited to simulating the generation of very short pulses (on the order of a picosecond or cavity round-trip time). However, this method is less suitable for the calculation of the IM and FM response, the harmonic distortion, the linewidth, and the RIN.

4.5 DERIVATION OF THE RATE EQUATIONS

In this section we want to explain the link between the coupled-wave equations of Chapter 3 and the lumped rate equations of Chapter 2 for a single-section laser. For ease of discussion, we will only consider one mode (with reference frequency ω_0) and neglect any spontaneous emission or noise. Moreover, gain suppression is assumed to be caused by the average photon density in the cavity.

We now consider the static field distribution $R_0^\pm(z)$ occurring at some arbitrarily chosen bias level. Consequently, the functions $R_0^\pm(z)$ are solutions of the coupled-wave equations with the complex Bragg deviation coefficient $\Delta\beta_0(z)$. Next we approximate the field of the laser with

$$R^\pm(z, t) = h(t)f^\pm(z)$$

$$f^\pm(z) = R_0^\pm(z) / \sqrt{\frac{1}{L}\int_0^L (|R_0^+|^2 + |R_0^-|^2)\, dz} \quad (4.26)$$

In this way the time and spatial variations are separated into two different functions. The time-dependent function $|h(t)|^2$ corresponds to the average power density in the laser. The z-dependent function describes the forward or backward field distribution normalized to a total intensity of 1. Inserting (4.30) into the wave equations then gives

$$\frac{1}{v_g} \frac{dh}{dt} + j\Delta\beta_L h = 0$$

$$\Delta\beta_L = \frac{1}{L} \int_0^L (\Delta\beta - \Delta\beta_0)(|f^+|^2 + |f^-|^2) \, dz \qquad (4.27)$$

Next, $h(t)$ is split into its modulus and phase according to

$$h(t) = v_g h v_0 \sqrt{S(t)} \exp\left(j \int^t \Delta\omega(\tau) \, d\tau\right) \qquad (4.28)$$

Here, $S(t)$ represents the average photon density in the cavity and $\Delta\omega$ the instantaneous frequency. Using (4.28), (4.27) can be rewritten as

$$\frac{1}{v_g} \frac{dS}{dt} - 2Im(\Delta\beta_L)S = 0$$

$$\Delta\omega = -Re(\Delta\beta_L) \qquad (4.29)$$

In format these equations correspond to the photon rate equation (2.3) and the phase equation (2.7) of Chapter 2. By choosing the laser threshold as bias point, $\Delta\beta_0$ becomes z-independent. Using the static field equations for $R_0^\pm(z)$ or for $f^\pm(z)$, one then finds for $2\,Im(\Delta\beta_0)$

$$2\,Im(\Delta\beta_0) = \frac{1}{L}\left[|f^+(L)|^2 - |f^-(L)|^2 + |f^-(0)|^2 - |f^+(0)|^2\right]$$

$$- \frac{2}{L} \int_0^L Re[\kappa_{FB} f^-(f^+)^* + \kappa_{BF} f^+(f^-)^*] \, dz \qquad (4.30)$$

The first term of the right-hand side of (4.30) describes the facet losses while the second term expresses the threshold gain reduction realized when gain coupling occurs. Indeed, this second term becomes zero for the index-coupling part of (κ_{FB}, κ_{BF}), but not for the gain-coupling part. For FP lasers this second term also vanishes and it can easily be shown that the first term is then reduced to $(1/2L)\ln(1/R_1R_2)$. Knowing the field distributions at threshold, (4.30) can be calculated.

For $2\,\text{Re}(\Delta\beta_0)$, an expression similar to (4.30) can be derived. Knowing the value of $2\,\text{Re}(\Delta\beta_0)$ is only relevant if the exact value of the instantaneous frequency is required instead of its relative variations.

Before we can fully compare the photon density and phase equations of Chapter 2 with (4.29), we also need to assume that the carrier density N remains uniform throughout the laser cavity. In that case, the longitudinal carrier rate equation is reduced immediately to the corresponding equation (2.2). The coefficient $\Delta\beta_L$ becomes z-independent and we get $\Delta\beta_L = \Delta\beta - \Delta\beta_0$.

Using the definition of $\Delta\beta$ of Section 3.4.4, we can now write the following rate equations for the photon density and the carrier density in the general case of a complex coupled DFB laser:

$$\frac{dS}{dt} = (G - \gamma)S + |\kappa_g v_g| f_{st} S$$

$$\frac{dN}{dt} = \frac{J}{qd} - (AN + BN^2 + CN^3) - GS - |\kappa_g v_g| f_{st} S$$

$$f_{st} = \frac{4}{L|\kappa_g|} \int_0^L \text{Re}[\kappa_g f^-(f^+)^*]\,dz \qquad (4.31)$$

We included the first term of the right-hand side of (4.30) in the total loss $\gamma = v_g(\alpha_{int} + \alpha_{facet})$, while the second term of (4.30) is retained explicitly. The factor f_{st} is called the *standing-wave factor* and the associated term only appears in the photon density rate equation if gain or loss coupling occurs. In the carrier rate equation, the standing-wave effect must be included in the case of gain gratings (obtained, for instance, via a corrugated active layer), but not in the case of loss gratings (obtained, for instance, via a corrugated lossy cladding layer). Section 3.7 gives some additional details on this issue.

We can now compare the photon rate equations (2.3) and the phase equations (2.7) with (4.31) and (4.29). Both sets of equations are identical in form, but in DFB-lasers the facet loss term is strongly dependent on the wavelength. Moreover, in complex coupled DFB lasers, the photon density rate equation contains an extra term that reflects the standing-wave effect. Loss gratings also show an additional extra average loss α_{extra} that can be included in α_{int} (see Section 4.1.1). The carrier rate equation (2.2) is compared with (4.31). The only difference occurs for gain coupling with a gain grating. In this case the carrier rate equation is extended with an additional stimulated recombination term due to the standing-wave effect.

4.6 LONGITUDINAL SPATIAL HOLE BURNING

Below threshold, the stimulated recombination term can be neglected in the longitudinal carrier rate equation. In that case, a uniform carrier injection current $J(z)$ causes a uniform carrier distribution $N(z)$ along the laser. However, once above threshold, an optical power distribution builds up and this power distribution is rarely uniform. The stimulated recombination will consequently cause a nonuniform carrier distribution. In other words, areas of strong optical field intensity will show a lower carrier density. This is called *spatial hole burning* and here it relates to the longitudinal direction.

The place along the laser where spatial hole burning occurs is also related to the strength of the coupling of the grating. Basically, three coupling strengths can be distinguished: $\kappa L < 1$ or undercoupling, $\kappa L \approx 1$ or critical coupling, and $\kappa L > 1$ or overcoupling. For overcoupled lasers, spatial holes tend to appear in the middle of the grating, and for undercoupled lasers spatial holes tend to appear at the edges of the grating. For critical coupling, the carrier distribution shows less pronounced holes.

In a DFB laser, a uniform carrier distribution along the laser corresponds to a uniform complex Bragg deviation $\Delta\beta = \beta_c - \beta_g$. This is the case at threshold, where no stimulated recombination is present. However, once above threshold, the carrier distribution will become nonuniform, causing a nonuniform Bragg deviation along the laser. This means that the distributed Bragg reflection will change with bias, which in turn causes the phase resonance and distributed reflector losses to change.

In the following chapters, we will discuss the influence of spatial hole burning on several laser characteristics. These important effects can be described very well by a longitudinal coupled-wave model, but not by the simple rate equations of Chapter 2. Improvement of the rate equations has been proposed in the literature [25,26] by adding a second carrier rate equation. The carrier density is then approximated by an expression of the form $N(z, t) = N_u(t) + N_s(t)f(z, t)$, in which N_u is the average carrier density in the cavity and N_s is the amplitude of the spatial hole burning distribution $f(z, t)$ with $\int f(z, t)\,dz = 0$ and $\int f^2(z, t)\,dz = 1$ (integration over the cavity length). In this way spatial hole burning effects in DFB lasers can also be approximated in a lumped rate equation model. However, these extended models must be treated with care. Usually they are only valid for small deviations with respect to one or more steady-state solutions obtained by means of a longitudinal model. Moreover, the parameters used in such extended lumped models heavily rely on longitudinal models for their derivation. In practice, it remains difficult to model accurately even the small-signal dynamic changes (in form and size) of the spatial distribution of the carriers and photons using only one extra variable. We will not elaborate on such approaches.

4.7 COUPLING COEFFICIENTS FOR DFB LASERS

The coupling coefficient of a grating is a key parameter describing the strength of the feedback of a grating. As explained in Chapter 10, the final step in creating the grating is often a grating etching process. Either chemical etchants or dry (reactive ion) etching techniques are used. Chemical etchants usually lead to sinusoidal or triangular shapes. Calculated results for the coupling coefficients of such gratings are given in [27]. In the case of dry etching, the gratings have a trapezoidal shape. Calculated results for such gratings are presented in [28].

In general, the coupling coefficients can be calculated by using (3.12a) to (3.12c). These equations are derived from the perturbation analysis of Chapter 3. Usually a y-independent multilayer thin-film waveguide is assumed, as shown in Figure 4.9. This means that the lateral coordinate y can be dropped in (3.12). The field distribution of the lowest order TE mode $\phi_0(x)$ then becomes the field distribution of an unperturbed thin-film waveguide without grating. To simplify the calculations, usually, as opposed to (3.12), the square of the refractive index of the waveguide, $n^2(x, z)$, is first expanded in a Fourier series and only then is weighting introduced by integrating the appropriate Fourier component across the normalized field intensity distribution $|\phi_0|^2$.

The choice of the proper unperturbed waveguide for the calculation of the coupling coefficients is not evident. The most accurate method is to replace the grating with an additional layer with an intermediate refractive index [29]. This approach, however, introduces an additional layer in the unperturbed thin-film waveguide. An easier approach, which shows adequate accuracy, is obtained by choosing the boundary of the unperturbed multilayer waveguide, as depicted in Figure 4.9 [27]. The boundary is chosen such that the shaded areas of the grating above and under the boundary of the unperturbed waveguide are equal. The appropriate eigenmode of the unperturbed waveguide can now be determined. General techniques include the *beam propagation method* [30] and the *staircase method* [31].

Once the field distribution of the unperturbed waveguide is known and the proper Fourier component of n^2 is calculated, the coupling coefficient can be

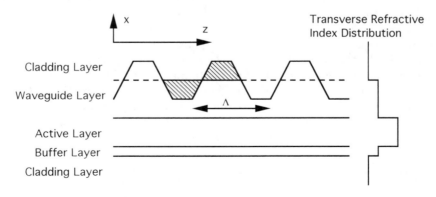

Figure 4.9 Corrugated waveguide structure. The waveguide with the dashed line is used as an unperturbed reference waveguide.

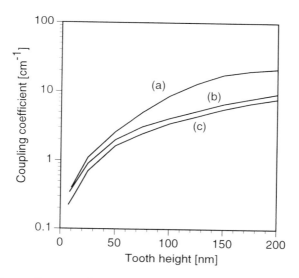

Figure 4.10 Coupling coefficient as a function of tooth height for: (a) a rectangular grating with duty cycle 0.25; (b) a symmetric triangular grating; and (c) a sawtooth grating.

determined. In [27,28], several numerical and analytic results are presented. Usually for rectangular, triangular, trapezoidal, and sinusoidal grating forms, complex analytic expressions for the coupling coefficients can be obtained. In general, one observes increasing coupling coefficients for increasing height of the teeth of the grating [27]. Figure 4.10 illustrates this for a rectangular grating.

The above calculation approach is valid for first-order gratings as well as second-order gratings. In each case the correct Fourier component of n^2 must be used.

References

[1] Björk, G., and O. Nilsson, "A New Exact and Efficient Numerical Matrix Theory of Complicated Laser Structures: Properties of Asymmetric Phase-Shifted DFB Lasers," *IEEE J. Lightw. Tech.*, Vol. LT-5, January 1987, pp. 140–146.
[2] Yamada, M., and K. Sakuda, "Analysis of Almost-Periodic Distributed Feedback Slab Waveguides via a Fundamental Matrix Approach," *Applied Optics*, Vol. 26, 15 August 1987, pp. 3474–3478.
[3] Buus, J., *Single Frequency Semiconductor Lasers*, SPIE Optical Engineering Press, 1990.
[4] Kogelnik, H., and C. V. Shank, "Coupled-Wave Theory of Distributed Feedback Lasers," *J. Appl. Phys.*, Vol. 43, No. 5, May 1972, pp. 2327–2335.
[5] David, K., J. Buus, G. Morthier, and R. Baets, "Coupling Coefficients in Gain Coupled DFB Lasers: Inherent Compromise Between Coupling Strength and Loss," *IEEE Phot. Tech. Lett.*, Vol. 3, May 1991, pp. 439–441.
[6] Yariv, A., and M. Nakamura, "Periodic Structures for Integrated Optics," *IEEE J. Quant. Electron.*, Vol. QE-13, No. 4, April 1977, pp. 233–253.

[7] Buus, J., "Mode Selectivity in DFB Lasers With Cleaved Facets," *Electron. Lett.*, Vol. 21, 1985, pp. 179–180.

[8] Mols, P., P. Kuindersma, W. Van Es, and I. Baele, "Yield and Device Characteristics of DFB Lasers: Statistics and Novel Coating Design in Theory and Experiment," *IEEE J. Quant. Electron.*, Vol. 25, June 1989.

[9] Whiteaway, J. E. A., G. H. B. Thompson, A. J. Collar, and C. J. Armistead, "The Design and Assessment of λ/4 Phase-Shifted DFB Laser Structures," *IEEE J. Quant. Electron.*, Vol. 25, No. 6, June 1989, pp. 1261–1279.

[10] Whiteaway, J. E. A., B. Garrett, G. H. B. Thompson, A. J. Collar, C. J. Armistead, and M. J. Fice, "The Static and Dynamic Characteristics of Single and Multiple Phase-Shifted DFB Laser Structures," *IEEE J. Quant. Electron.*, Vol. 28, No. 5, May 1992, pp. 1277–1293.

[11] Kazarinov, R., and C. Henry, "Second-Order Distributed Feedback Lasers With Mode Selection Provided by First-Order Radiation Losses," *IEEE J. Quant. Electron.*, Vol. QE-21, No. 2, February 1985, pp. 144–150.

[12] Makino, T., and J. Glinski, "Effects of Radiation Loss on the Performance of Second-Order DFB Semiconductor Laser," *IEEE J. Quant. Electron.*, Vol. QE-24, No. 1, January 1988, pp. 73–82.

[13] Morthier, G., and R. Baets, "Modelling of Distributed Feedback Lasers," Chapter 7 in *Compound Semiconductor Device Modelling*, C. M. Snowden and R. E. Miles, eds., London: Springer-Verlag, 1993, pp. 119–148.

[14] Vankwikelberge, P., G. Morthier, and R. Baets, "CLADISS, a Longitudinal, Multi-Mode Model for the Analysis of the Static, Dynamic and Stochastic Behavior of Diode Lasers With Distributed Feedback," *IEEE J. Quant. Electron.*, October 1990.

[15] Photon Design, IMEC/University of Gent, i-CLADISS User Manual, Photon Design, Oxford, 1995.

[16] Agrawal, G. P., and A. H. Bobeck, "Modeling of Distributed Feedback Semiconductor Lasers With Axially-Varying Parameters," *IEEE J. Quant. Electron.*, Vol. 24, No. 12, December 1988, pp. 2407–2413.

[17] Makino, T., "Transfer-Matrix Theory of the Modulation and Noise of Multi-Element Semiconductor Lasers," *IEEE J. Quant. Electron.*, Vol. 29, No. 11, November 1993, pp. 2762–2770.

[18] Olesen, H., B. Tromborg, X. Pan, and H. E. Lassen, "Stability and Dynamic Properties of Multi-Electrode Laser Diodes Using a Green's Function Approach," *IEEE J. Quant. Electron.*, Vol. 29, No. 8, August 1993, pp. 2282–2301.

[19] Takemoto, A., H. Watanabe, Y. Nakajima, Y. Sakakibara, S. Kakimoto, and H. Namizaki, "Low Harmonic Distortion Distributed Feedback Laser Diode and Module for CATV Systems," *Proc. Opt. Fiber Communications Conf. (OFC'90)*, San Francisco, 1990, p. 213.

[20] Lau, K., and A. Yariv, "Intermodulation Distortion in Semiconductor Injection Lasers," *Appl. Phys. Lett.*, Vol. 45, 1984, pp. 1034–1036.

[21] Chen, Y., H. Winful, and J. Liu, "Subharmonic Bifurcations and Irregular Pulsing Behavior of Modulated Semiconductor Lasers," *Appl. Phys. Lett.*, Vol. 47, August 1985, pp. 208–210.

[22] Vahala, K., and A. Yariv, "Semiclassical Theory of Noise in Semiconductor Lasers—Part I," *IEEE J. Quant. Electron.*, Vol. 19, June 1983, pp. 1096–1101.

[23] Lowery, A. J., "Dynamic Modelling of Distributed Feedback Lasers Using Scattering Matrices," *Electron. Lett.*, Vol. 25, No. 19, 14 September 1989, pp. 1307–1308.

[24] Zhang, L. M., S. F. Yu, M. C. Nowell, D. D. Marcenac, J. E. Carroll, and R. G. Plumb, "Dynamic Analysis of Radiation and Side-Mode Suppression in a Second-Order DFB Laser Using Time-Domain Large-Signal Traveling Wave Model," *IEEE J. Quant. Electron.*, Vol. 30, No. 6, June 1994, pp. 1389–1395.

[25] Kinoshita, J., "Modelling of High-Speed DFB Lasers Considering the Spatial Holeburning Effect Using Three Rate Equations," *IEEE J. Quant. Electron.*, Vol. 30, No. 4, April 1994, pp. 929–938.

[26] Schatz, R., "Dynamics of Spatial Hole Burning Effects in DFB Lasers," *IEEE J. Quant. Electron.*, Vol. 31, No. 11, November 1995, pp. 1981–1993.

[27] Streifer, W., D. R. Scifres, and R. Burnham, "Coupling Coefficients for Distributed Feedback Single- and Double-Heterostructure Diode Lasers," *IEEE J. Quant. Electron.*, Vol. QE-11, No. 11, November 1975, pp. 867–873.

[28] Correc, P., "Coupling Coefficients for Trapezoidal Gratings," *IEEE J. Quant. Electron.*, Vol. QE-24, No. 1, January 1988, pp. 8–10.

[29] Handa, K., S. T. Peng, and T. Tamir, "Improved Perturbation Analysis of Dielectric Gratings," *Appl. Phys.*, Vol. 5, 1975, pp. 325–328.

[30] Van Roey, J., J. Van der Donk, and P. E. Lagasse, "Beam Propagation Method: Analysis and Assessment," *J. Opt. Soc. of America*, Vol. 71, No. 3, 1981, pp. 803–810.

[31] Vankwikelberge, P., J. P. Van de Capelle, R. Baets, B. H. Verbeek, and J. Opschoor, "Local Normal Mode Analysis of Index-Guided AlGaAs Lasers With Mode Filter," *IEEE J. Quant. Electron.*, Vol. QE-23, No. 6, June 1987, pp. 730–737.

Chapter 5

A Closer Look at the Carrier Injection

In the analysis of laser diodes, it is often assumed that a spatially uniform current density is injected into the active layer, where all injected carriers recombine. In reality, there is a uniform voltage on the electrodes, the current is not completely directed toward the active layer, and not even all carriers created in the active layer recombine there. Consequently, an electrical representation of the lateral/transverse structure does not consist of a current source in series with a diode, but of a voltage source that drives a more complex circuit. A diode in series with a resistance can be distinguished in this circuit, but different parasitic elements such as capacitances, transistors, or thyristors, which cause current leakage (i.e., current not reaching the active layer), are also connected with the source. A typical circuit representation of the lateral/transverse structure is given in Figure 5.1 (it actually corresponds to a double-channel planar buried heterostructure (DCPBH) laser [1]). It will become clear further on that the values of the circuit elements will depend on the biasing of the laser and vary along the longitudinal direction.

Leakage currents and parasitics must be reduced as much as possible, since they can cause the behavior to deteriorate in several respects. Leakage currents increase the threshold current and reduce the efficiency. Therefore, they also give rise to increased heating, which limits the achievable output power and the achievable tuning range in tunable lasers. In addition, they were shown to be a cause of harmonic distortion under analog modulation (e.g., for CATV applications) [2].

Passive parasitic elements such as capacitances and resistances are mainly of importance for the dynamic behavior and may limit the achievable modulation bandwidth. Well-known parasitic elements include the series resistances of the cladding layers, the capacitances associated with current blocking layers, diffusion capacitances, and depletion capacitances. Recently it has been pointed out that the carrier transport times (e.g., across separate confinement layers) in quantum-well structures and the propagation of voltage and current waves along the electrodes (whereby the

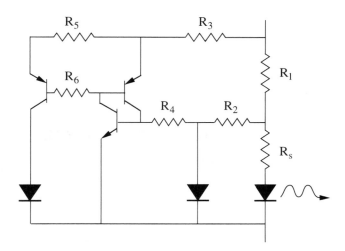

Figure 5.1 Equivalent circuit for a DCPBH laser. (*After:* [1].)

laser structure can be regarded as a stripline) should also be taken into account at high (microwave) frequencies [3,4].

The different causes of carrier leakage and their representation by parasitic elements in an equivalent electronic circuit are the subject of this chapter. This chapter relies partly on theory from semiconductor physics, but more specialized topics such as heterojunctions and semi-insulating materials are first briefly introduced. Sections 5.1 to 5.4 focus on current leakage under static operation, while Sections 5.5 and 5.6 discuss parasitic capacitances and microwave effects. In Section 5.7 we illustrate possible modeling tools to deal with the electronics of carrier injection.

Current injection, or, more generally, the electronic behavior, of the lateral/transverse geometry is not the easiest aspect of laser behavior, partly because many different structures are used. We nevertheless hope that this chapter can provide some insight into the complex physics.

5.1 HETEROJUNCTIONS AND SEMI-INSULATING MATERIALS

5.1.1 Heterojunctions

Lateral/transverse laser diode structures always comprise one or more heterojunctions (i.e., junctions formed by two intrinsically different semiconductor materials). Such junctions are basically different from homojunctions (two intrinsically identical semiconductors in thermodynamic equilibrium) because of the difference in bandgap of the two materials. The diffusion voltage is therefore no longer solely determined by the doping levels, but contains an additional contribution from the bandgap difference. If no graded junctions are considered, the static electronic behavior in the homogeneous regions is governed by the following semiconductor equations for the potential V and current densities $\mathbf{J}_n$ (for electrons) and $\mathbf{J}_p$ (for holes) [5]:

$$\nabla \cdot (\varepsilon \nabla V) = q(N - P + N_A^- - N_D^+)$$

$$\nabla \cdot \mathbf{J}_p + q(R - G) = 0$$

$$\nabla \cdot \mathbf{J}_n - q(R - G) = 0 \qquad (5.1)$$

combined with the following constitutive relations:

$$\mathbf{J}_p = P\mu_p \nabla F_p, \ \mathbf{J}_n = N\mu_n \nabla F_n$$

$$P = N_v F_{1/2}\left(\frac{F_p - E_v}{kT}\right), \ N = N_c F_{1/2}\left(\frac{F_n - E_c}{kT}\right) \qquad (5.2)$$

N and P are the densities of electrons and holes, respectively. They have mobilities μ_n and μ_p, respectively, and quasi-Fermi levels F_n and F_p, respectively. N_c and N_v are the effective density of states of conduction and valence band, with band edges E_c and E_v, respectively. ε is the dielectric constant and $R - G$ is the net recombination rate, which can be expressed as a function of the carrier densities. N_A and N_D are the acceptor and donor concentration. We will further always assume that all dopants are ionized (this is the case at temperatures near room temperature). $F_{1/2}$ is the Fermi-Dirac integral

$$F_{1/2}(F) = \frac{2}{\sqrt{\pi}} \int_0^\infty \frac{E^{1/2}}{1 + \exp(E - F)} \, dE \qquad (5.3)$$

The band edges are affected by the potential V according to

$$E_{c,v} = E_{c,v0} - qV \qquad (5.4)$$

Table 5.1 gives an overview of the material parameters that define the electronic behavior of InP and 1.3- and 1.55-μm InGaAsP [6].

The electric field and potential at the double heterojunction formed by the active layer can be easily calculated assuming abrupt space-charge regions in the cladding layers. As illustrated in Figure 5.2, this assumption is valid if the width of the space-charge regions is large compared with the Debye or screening length. It leads to linearly varying electric fields and quadratic potentials in the cladding layers

Table 5.1
Main Material Properties of InP and 1.55-μm InGaAsP

Parameter (unit)	InP	InGaAsP, 1.55	InGaAsP, 1.3
Bandgap E_g (eV)	1.344	0.8	0.96
Electron mobility μ_n (cm2/Vs)	2,300	9,000	5,000
Hole mobility μ_p (cm^2/Vs)	95	160	120
Diffusion constant Dn (cm^2/s)	59	230	128
Diffusion constant D_p (cm^2/s)	2.43	4.096	3.072
Eff. density of states C.B. (cm^{-3})	5.8×10^{17}	2.65×10^{17}	3.65×10^{17}
Eff. density of states V.B. (cm^{-3})	1.08×10^{19}	7×10^{18}	7.1×10^{18}
Carrier lifetime (ns)	5	5	5

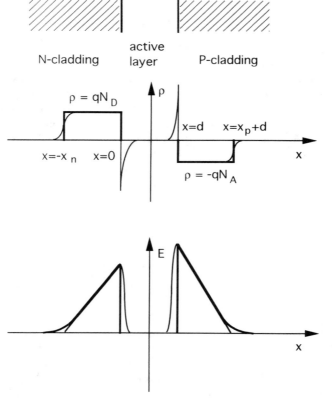

Figure 5.2 Charge density and electric field at a double heterojunction. Exact profiles and abrupt depletion layer approximation are shown by thin and thick lines, respectively.

$$E(x) = \frac{qN_D}{\varepsilon}(x + x_n) \quad \text{for } -x_n < x < 0$$

$$E(x) = \frac{qN_A}{\varepsilon}(x - d - x_p) \quad \text{for } d < x < d + x_p \tag{5.5}$$

In the active region, where we assume that the carrier density is much larger than the doping levels and hence $N = P$, the potential can be found after solution of Poisson's equation

$$\frac{d^2V}{dx^2} = \frac{q}{\varepsilon}(N - P) = \frac{q}{\varepsilon}\frac{\partial(N - P)}{\partial V}V = \frac{1}{L_s^2}V \tag{5.6}$$

in which L_s is the screening length. The solutions of this equation are exponentially decaying potentials and hence exponentially decaying fields. Assuming an identical dielectric constant in cladding and active layers and expressing the continuity of the electric field results in the following fields inside the active layer

$$E(x) = \frac{qN_D}{\varepsilon} x_n \exp\left(-\frac{x}{L_s}\right) \quad \text{for } 0 < x$$

$$E(x) = \frac{qN_A}{\varepsilon} x_p \exp\left(\frac{x - d}{L_s}\right) \quad \text{for } x < d \tag{5.7}$$

The derivation of the potential $V(x)$ is straightforward and one finds

$$V(x) = -\frac{qN_D}{2\varepsilon}(x + x_n)^2 \quad \text{for } -x_n < x < 0$$

$$V(x) = -\frac{qN_D}{2\varepsilon}\left[x_n^2 + 2x_n L_s - 2x_n L_s \exp\left(-\frac{x}{L_s}\right)\right] \quad \text{for } 0 < x < d/2$$

$$V(x) = -\frac{qN_D}{2\varepsilon}[x_n^2 + 2x_n L_s] - \frac{qN_A}{2\varepsilon} 2x_p L_s \exp\left(\frac{x + d}{L_s}\right) \quad \text{for } d/2 < x < d$$

$$V(x) = -\frac{qN_D}{2\varepsilon}[x_n^2 + 2x_n L_s] - \frac{qN_A}{2\varepsilon}[2x_p L_s + (x - d)^2] \quad \text{for } d < x < d + x_p \tag{5.8}$$

The widths x_p and x_n can then be determined from the carrier concentrations (or Fermi levels) on both sides of the two junctions. For example, expressing the electron concentrations at $x = -x_n$ (where $N = N_D$) and at $x = d/2$ (where $N = N_a$) and making use of the constant quasi-Fermi level F_n in that region gives

$$\frac{N_D}{N_{ccl}} = F_{1/2}\left(\frac{E_{ccl} - F_n}{kT}\right) \text{ and } \frac{N_a}{N_{ca}} = F_{1/2}\left(\frac{E_{ca} - F_n}{kT}\right) \tag{5.9}$$

with E_{ccl} and E_{ca} the conduction band edges in cladding, respectively, active layer and N_{ccl} and N_{ca} the effective density of states of the conduction band in cladding and active layer. The difference in potential between $x = -x_n$ and $x = d/2$ can be used to relate E_{ccl} to E_{ca}

$$E_{ccl} = E_{ca} + \Delta E_c - \frac{q^2 N_D^+}{2\varepsilon} x_n (x_n + 2L_s) \tag{5.10}$$

Inversion of the relations (5.9) then results in an explicit relation between x_n and the carrier densities

$$\Delta E_c - \frac{q^2 N_D^+}{2\varepsilon} x_n(x_n + 2L_s) = kT\left[F_{1/2}^{-1}\left(\frac{N_D^+}{N_{ccl}}\right) - F_{1/2}^{-1}\left(\frac{N_a}{N_{ca}}\right)\right] \tag{5.11}$$

A similar relation can be used to determine x_p. A good approximation for the inverse functions has been given by Joyce and Dixon [7]

$$F_{1/2}^{-1}(y) = \ln(y) + \frac{\sqrt{2}}{4} y + \left(\frac{3}{16} - \frac{\sqrt{3}}{9}\right)y^2 \tag{5.12}$$

Note that the screening length is very small (on the order of a few nanometers) for the doping and carrier densities existing in laser diodes above threshold [8].

5.1.2 Semi-Insulating InP

In order to achieve a considerable reduction of current leakage and the influence of parasitics, people increasingly use semi-insulating InP (SI-InP). The background shallow donor concentration in this material is compensated for by deep acceptor traps created through Fe doping. SI-InP has intrinsic behavior in addition to high-resistivity properties [9]. The steady-state trap filling f of the deep acceptors can be described by a Shockley-Read-Hall model

$$f = \frac{\tau_{cp} N + \tau_{cn} P}{\tau_{cp}(N + N_0) + \tau_{cn}(P + P_0)} \tag{5.13}$$

in which τ_{cn} and τ_{cp} are the time constants for capture and emission of electrons and holes, respectively, and N_0 and P_0 are the electron and hole concentrations when the Fermi level is at the same energy as the trap

$$N_0 = N_i \exp\left(\frac{E_T - E_i}{kT}\right), \quad P_0 = \frac{N_i^2}{N_0} \tag{5.14}$$

with E_i the Fermi level of intrinsic material. If only N-type doping is present, the trap filling reduces to

$$f = \frac{N}{N + N_0} \qquad (5.15)$$

This relation can then be used to derive the charge density caused by the traps. Donor traps are neutral when filled and positive when empty, while acceptor traps are negative when filled and neutral when empty. For SI-InP, Poisson's equation takes the following form

$$\nabla^2 V = -\frac{q}{\varepsilon}[N_D^+ - N + P - N_{Ad}f] \qquad (5.16)$$

where N_{Ad} is the concentration of deep acceptors. The concentration of deep acceptors required to have intrinsic material (with $N = P = N_i$) can be derived from the neutrality condition and is simply $N_D/f(N_i)$.

The I-V characteristics for a layer of thickness d of semi-insulating material between two perfect N-type contacts are depicted in Figure 5.3 [8]. Diffusion has been neglected in the derivation of the characteristics. The quantities of the x- and y-axis are defined as follows:

$$\frac{1}{w_a} = \frac{J}{\left(\dfrac{q^2 N_D^2 \mu_n d}{\varepsilon}\right)}$$

$$\frac{v_a}{w_a^2} = \frac{V}{\left(\dfrac{q N_D d^2}{\varepsilon}\right)} \qquad (5.17)$$

With these variables, the characteristics are independent of the dielectric constant and thickness d. The parameter A used in the figure is defined as

$$A = \frac{N_{Ad} N_0}{N_D^2} \qquad (5.18)$$

The behavior is governed by Ohm's law as long as the traps are little filled. Indeed, the current is caused by drift of the thermally generated electrons and holes (with fixed concentration independent of the voltage), since all injected carriers are trapped.

A perfect isolator, with the current being proportional to the square of the voltage, results if the traps are completely filled. The current is now caused by the drift of injected electrons in the conduction band. However, all the injected carriers contribute to a free space charge and hence an electric field. In other words, there is no charge neutrality as in semiconductors, where the concentration of ionized impurities equals the free carrier concentration. Therefore, both the drift velocity and the carrier concentration are proportional to the voltage. Important is the voltage at which the transition between both regimes occurs

$$V_t = \frac{qN_{Ad}N_0 d^2}{2\varepsilon N_D} \tag{5.19}$$

As can be seen in Figure 5.3, the current is very low below V_t and increases over several orders of magnitude around this voltage.

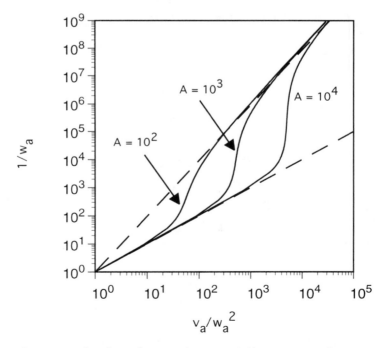

Figure 5.3 I-V characteristics for a layer of semi-insulating material between two perfect contacts (the dotted line is an approximate solution). (*After:* [9].)

5.2 CARRIER LEAKAGE OVER HETEROBARRIERS

The leakage in laser diodes generally consists of two parts: the carriers leaking over the heterobarriers (e.g., re-escaping from the active layer) and current flow along paths not reaching the double heterojunction (Figure 5.4). It has been mainly carrier leakage over the heterobarriers that has been discussed in the literature so far [1,10]. Due to the specific form of the active layer (wide in the lateral direction and narrow in the transverse direction), this carrier leakage occurs predominantly in the transverse direction.

We will derive the leakage of electrons into the P-cladding layer as an example (Figure 5.5). With N_{ccl} the effective density of states of the conduction band in the cladding layer and F_n the Fermi level for the electrons in the active layer (with respect to the conduction band edge of the active layer and positive if the Fermi level is inside the conduction band), one finds for the electron density N_b at the i-P heterojunction

$$N_b = N_{ccl} \exp\left(-\frac{\Delta E_c}{kT}\right) \exp\left(\frac{F_n}{kT}\right) \quad (5.20)$$

with ΔE_c the discontinuity of the conduction band at the heterojunction. The electron current in the P-cladding can then be calculated from the current continuity equation. When both drift and diffusion are taken into account, one can write

$$\frac{\partial^2 N}{\partial x^2} - \frac{\mu_n E}{D_n} \frac{\partial N}{\partial x} - \frac{N - N_0}{L_n^2} = 0 \quad (5.21)$$

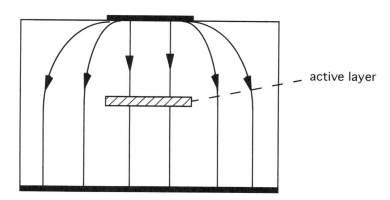

Figure 5.4 Laser diode and the different possible current paths.

with D_n the electron diffusion constant and L_n the diffusion length. A constant electric field E has been assumed in (5.21), which means that an area outside the space-charge region is considered. The solution of the modified diffusion equation (5.21) can be expressed as

$$J_{L,N} = qD_n N_b \left[\sqrt{\frac{1}{L_n^2} + \frac{1}{4z^2}} \coth\left(\sqrt{\frac{1}{L_n^2} + \frac{1}{4z^2}}\, d_P\right) + \frac{1}{2z}\right]$$

$$z = \frac{D_n}{\mu_n E} = \frac{kT}{q}\frac{\sigma_P}{J_{\text{tot}}} \qquad (5.22)$$

with d_p the thickness of the cladding layer. J_{tot} is the total injected current and σ_P the conductivity of the P-layer. A similar expression can be derived for the holes leaking into the N-cladding. However, the hole leakage is always considerably smaller than the electron leakage because of the smaller mobility of the holes.

The expression for the transverse leakage current decreases for increasing value of z and thus for increasing conductivity or doping of the cladding layers. In the limit of very large doping and small (with respect to the diffusion length, which is typically 5 µm) thickness of the P-layer, one finds the leakage current

$$J_{L,N} = \frac{qD_n N_b}{d_P} \qquad (5.23)$$

which decreases with increasing thickness of the cladding layer. For large thicknesses of the P-layer, the leakage current approximates the value $J_{L,N} = qD_n N_b/L_n$.

It can be seen that the discontinuity in the conduction band itself depends on the doping level. Indeed, the value N_b is the electron density at the boundary of the space-charge region and, as can be seen from Figure 5.5, the difference in conduction band edge at this point with the active layer is affected by the depletion field. When

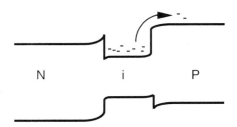

Figure 5.5 Double heterojunction formed by the active layer (intrinsic) and the cladding layers in the transverse direction of a laser diode.

the screening length in the active layer is neglected with respect to the width of the depletion layer, one finds for N_b:

$$N_b = N_{ccl} \exp\left(-\frac{\Delta E_c}{kT}\right) \exp\left(\frac{F_n}{kT}\right) \frac{P_a}{N_A^-} \frac{N_{vcl}}{N_{va}} \tag{5.24}$$

with P_a the hole density in the active layer and N_A the doping level of the P-cladding. N_{va} and N_{vcl} are the effective density of states of the valence band in the active and cladding layers, respectively. (Note that N_{vcl} is usually much larger than N_{va}). From (5.24) one can also see an additional decrease of N_b (and thus of the leakage) with increasing doping level. F_n obviously depends on the electron density N_a in the active layer (and $N_a = P_a$ for small doping levels). Figure 5.6 shows calculated values of N_b as a function of the carrier density in the active layer for three different doping levels and for a 1.55-µm bulk InGaAsP active layer. It can be noted that because of the exponential dependence on the bandgap difference, the values of N_b and thus of the leakage are about a factor of 500 larger for a 1.3-µm bulk InGaAsP active layer.

The current leakage over the heterobarriers is in general very small (and less than 1% of the injected current) and may only be of some importance for a 1.3-µm active layer in the case of weak doping levels, high temperatures, and high injection.

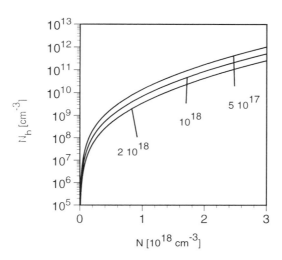

Figure 5.6 Carrier density outside the active layer vs. active layer carrier density for different doping levels (in cm^{-3}).

5.3 CARRIER INJECTION IN GAIN-GUIDED AND WEAKLY INDEX-GUIDED LASERS

Gain-guided structures (in which the waveguiding is entirely caused by the higher gain in a part of the active layer) and weakly index-guided structures (in which the waveguiding can be partly attributed to the higher gain in a part of the active layer) are little used in optical communication lasers. They nevertheless possess a rather simple and easy-to-grow structure and are still used for short-wavelength lasers. The current injection mechanisms in these structures are easily described and hence ideal to start a description of current injection. Some typical structures (the oxide stripe laser and the ridge waveguide laser) are shown in Figure 5.7.

The carrier injection in such structures can be described analytically if the passive P-layer under the oxide is relatively thin and infinitely broad [1]. With $I(y)$ the lateral current per unit distance in the longitudinal direction, $V(y)$ the potential over the active layer, and $J(y)$ the transverse current density (Figure 5.8), one has the relations

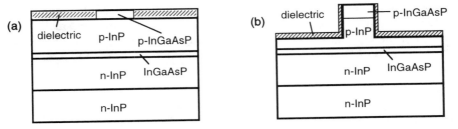

Figure 5.7 (a) The oxide stripe laser and (b) the ridge waveguide laser.

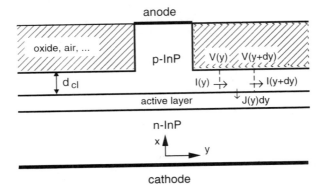

Figure 5.8 Current injection in an oxide stripe laser.

A Closer Look at the Carrier Injection

$$Jdy = I(y) - I(y + dy) \quad \text{or} \quad J = -\frac{dI}{dy}$$

$$V(y) - V(y + dy) = \frac{\rho}{d_{cl}} I dy \quad \text{or} \quad I = -\frac{d_{cl}}{\rho} \frac{dV}{dy} \quad (5.25)$$

with d_{cl} the thickness of the passive layer under the oxide and ρ the resistance of this passive layer. For $J(y)$ one can assume the usual diode relation

$$J(y) = J_S \exp\left(\frac{qV(y)}{nkT}\right) \quad (5.26)$$

with n the ideality factor of the diode. J_S is the saturation current density

$$J_S = qN_i^2 \left(\sqrt{\frac{D_p}{\tau_N}} \frac{1}{N_D} + \sqrt{\frac{D_n}{\tau_p}} \frac{1}{N_A} \right) \quad (5.27)$$

with τ_N and τ_P the carrier lifetimes in the N- and P-cladding, respectively. Derivation of $J(y)$ with respect to y then gives

$$\frac{d^2I}{dy^2} = -\frac{dJ}{dy} = -\frac{q}{nkT} \frac{dV}{dy} J(y) = -\frac{q\rho}{nkTd_{cl}} \frac{dI}{dy} I \quad (5.28)$$

with the boundary condition that $I \to 0$ if $y \to \pm\infty$. The solution of this equation is [11]

$$J(y) = J_0 \quad \text{for} \quad |y| < \frac{w}{2}$$

$$J(y) = \frac{J_0}{\left(1 + \frac{|y| - \frac{w}{2}}{l_0}\right)^2} \quad \text{for} \quad |y| > \frac{w}{2}$$

$$(5.29)$$

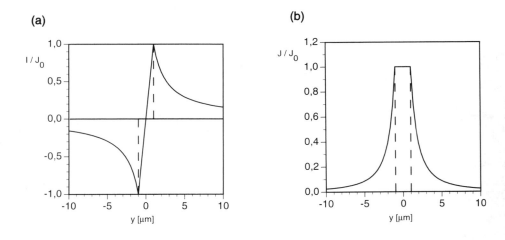

Figure 5.9 Current spreading in the passive top layer: (a) lateral current and (b) transverse current.

whereby it is assumed that the transverse current density under the contact is uniform. J_0 is determined by the total injected current and I_0 is given by

$$I_0 = \frac{2nkTd_{cl}}{q\rho J_0} \tag{5.30}$$

The functions I and J are depicted in Figure 5.9 for $J_0 = 3 \times 10^{-5}$ A/µm², $w = 2$ µm, and $l_0 = 1.67$ µm.

It can be seen that the previous derivation only holds if the stimulated emission is still small and for small current levels. Stimulated emission, which is not uniform because of the nonuniform optical intensity, would affect the carrier recombination times and cause a nonuniform saturation current density J_S.

5.4 LATERAL CURRENT LEAKAGE IN INDEX-GUIDED STRUCTURES

Two commonly used index-guided structures, the DCPBH and the etched-mesa buried heterostructure (EMBH) are shown in Figure 5.10.

The leakage currents in EMBH and related structures (e.g., planar buried heterostructures and strip buried heterostructures) can be analyzed qualitatively by considering a structure as shown in Figure 5.11. One can recognize the following leakage

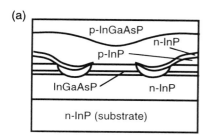 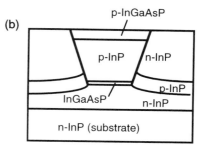

Figure 5.10 (a) DCPBH and (b) EMBH structures.

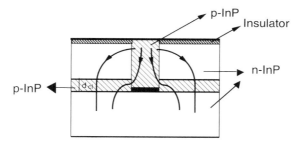

Figure 5.11 Scheme of an EMBH structure and the leakage paths.

paths in this structure: a forward-biased p-n junction and a p-n-p-n junction or thyristor. From the analysis in Section 5.3 it follows that the current density flowing through the p-n junction as a function of the distance y from the active layer can be approximated as

$$J(y) = \frac{J_0}{\left(1 + \frac{y}{I_0}\right)^2} \quad \text{with} \quad I_0 = \frac{2nkTd_{cl}}{q\rho J_0} \tag{5.31}$$

For the total leakage current, one then finds

$$I_{L,1} = 2LJ_0 = 2L\sqrt{\frac{2nkTd_{cl}J_S}{q\rho}} \exp\left(\frac{qV}{2nkT}\right) \qquad (5.32)$$

with V the voltage at $y = 0$, which equals the voltage drop over the active layer. It can also be seen that the current from the top P-layer can only reach the P-layer adjacent to the active layer over a thickness $d_{cl} - d$, so that in (5.32) d_{cl} should be replaced by $d_{cl} - d$ (with d the thickness of the active layer) or more generally a number between d_{cl} and $d_{cl} - d$.

The current $I_{L,1}$ acts as base current for an n-p-n transistor, and the total emitter current (which is the total lateral leakage current) of that transistor can be approximated by

$$I_L = I_{L,1} \frac{1}{1 - \alpha_{tr}} \quad \text{with} \quad \alpha_{tr} \approx \left[\cosh\left(\frac{d_{cl}}{L_n}\right)\right]^{-1} \qquad (5.33)$$

with α_{tr} the α-factor of the transistor. The expression given for this α-factor is the one holding for a cylindrical transistor and it will be slightly different for the geometry of Figure 5.11. This factor also depends somewhat on the applied voltages, because of the so-called Early effect, and will increase with increasing current injection.

From (5.32) and (5.33) one can conclude that the lateral leakage mainly depends on the temperature, the carrier lifetimes, the thickness d_{cl}, and the resistivity ρ of the layer adjacent to the active layer. The effect of the doping level in the adjacent P-layers, as it follows from numerical simulations, is shown in Figure 5.12 for a structure with a total thickness of 5 µm and a width of 10 µm, with 3-µm-thick N-cladding, $d_{cl} = 0.5$ µm, $d = 0.2$ µm, and a 2-µm-wide active layer. The nominal doping is 5×10^{17} cm^{-3} for the N-layers and 10^{18} cm^{-3} for the P-layers. The advantage of a small doping level can be clearly seen, although the advantage seems to become smaller at higher current injection. The doping level N_A further affects the saturation current density J_S, but only partly, as can be seen from (5.27), while the resistivity is proportional to the inverse of the doping level. Figure 5.13 shows the influence of the thickness d_{cl} of the adjacent P-layers, which influences both the base current and the α-factor of the transistor.

Other types of index-guided structures are those with SI-InP. As an example, the commonly used semi-insulating buried heterostructure (SIBH) structure is shown in Figure 5.14. The leakage in such structures remains small if the voltage between the P- and N-claddings remains below V_t (given by (5.19)). This voltage is proportional

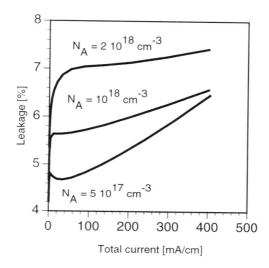

Figure 5.12 Influence of the doping level of the adjacent P-layers on the lateral leakage current for an EMBH structure.

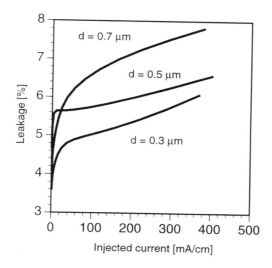

Figure 5.13 Influence of the thickness of the adjacent P-layer on the lateral leakage current in an EMBH laser.

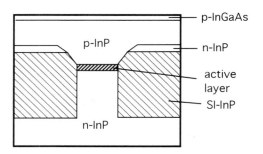

Figure 5.14 The SIBH structure.

to the squared thickness d of the SI-InP layer, while the resistance below this voltage is proportional to the thickness. Hence, the SI-InP layer should be rather thick to minimize the leakage currents.

5.5 PARASITIC ELEMENTS

The parasitic elements associated with the transverse current path consist of a series resistance and capacitances formed at the heterojunctions. Obviously, the value of these elements strongly depends on the doping levels of the cladding layers. The series resistance R_s, for example, is simply given by

$$R_s = \frac{1}{qwL}\left(\frac{d_N}{\mu_N N_D} + \frac{d_P}{\mu_P N_A}\right) \quad (5.34)$$

with w and L the width and length, respectively, of the active layer. A typical value for the series resistance is 2 Ω – 10Ω. The heterojunction capacitances consist of a depletion capacitance C_d (having its origin in the space-charge layers) and of a diffusion capacitance C_D. The depletion capacitance can be calculated from the relationship between space charges and potential over the junctions. For the junction between active layer and P-cladding, for example, one finds the differential capacitance

$$C_d = \frac{\varepsilon w L}{\sqrt{\dfrac{2\varepsilon}{q^2 N_A}\left\{\Delta E_r + kT\ln\left(\dfrac{N_A}{N_{rcl}}\right) - kT\ln\left(\dfrac{P_a}{N_{ra}}\right)\right\}}} \quad (5.35)$$

and a similar expression holds for the capacitance of the junction between active layer and N-cladding. A typical value for the total depletion capacitance of the active layer is a few picofarads. This value results in an R-C constant of 1 to 10 ps and therefore has as good as no influence on the dynamic characteristics. The diffusion capacitance of the heterojunctions has no influence, since the current consists practically completely of conduction of majority carriers (and not of diffusion of minority carriers as in homojunction diodes).

More important parasitic elements are, however, associated with the lateral leakage. Many index-guided lasers are structurally similar to the EMBH lasers and can be represented by an equivalent circuit, as shown in Figure 5.15, in which the passive region adjacent to the active layer and the homojunction are modeled as an R-C ladder network (with δR and δC infinitely small). It is easily derived (by expressing that the impedance of a chain of n $\delta R \delta C$ combinations equals that of a chain of $n + 1$ $\delta R \delta C$ combinations for $n \to \infty$) that the input impedance of this ladder network equals

$$Z_{in} = \frac{j\omega\delta R\delta C + \sqrt{(j\omega\delta R\delta C)^2 + 4j\omega\delta R\delta C}}{2j\omega\delta C} = \sqrt{\frac{\delta R}{j\omega\delta C}} = \sqrt{\frac{R}{j\omega C}} \qquad (5.36)$$

which results in the following cutoff frequency (with $R = \Sigma\delta R$ and $C = \Sigma\delta C$)

$$\omega_c = \frac{R}{R_s^2 C} \qquad (5.37)$$

From this expression, it follows that the cutoff frequency increases with decreasing resistance (and hence increasing doping level) of the cladding layers below and above the active layer, with decreasing capacitance of the homojunction and with increasing resistance of the cladding layer adjacent to the active layer. The junction capacitance

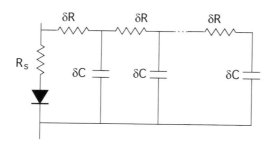

Figure 5.15 Equivalent circuit for a laser with distributed capacitance.

now also includes the diffusion capacitance. With the assumption of a negligible electric field and a narrow P-layer adjacent to the active layer, this capacitance can be expressed as [5]

$$C_D = \frac{q^2 N_i^2}{2kT} \frac{L_p}{N_D} \exp\left(\frac{qV}{kT}\right) w_{cl} L \tag{5.38}$$

with w_{cl} the total width of the cladding layer adjacent to the active layer and V the voltage over the p-n junction. This voltage is approximately equal to the voltage over the double heterojunction (and thus slightly larger than the bandgap voltage of the active layer material (e.g., 0.8V for 1.55-µm InGaAsP)). Substituting typical values for the different parameters in expression (5.38) indicates that the diffusion capacitance is very small and on the order of 1 to 100 fF. The depletion capacitance of the homojunction is given by

$$C = \sqrt{\frac{q\varepsilon N_A N_D}{2(N_A + N_D)}} \frac{w_{cl} L}{\sqrt{V_{bi} - V_g}} \tag{5.39}$$

with V_{bi} the built-in voltage

$$V_{bi} = \frac{kT}{q} \ln\left(\frac{N_A N_D}{N_i^2}\right) \tag{5.40}$$

With the assumption that N_D is very large, one can optimize the doping level of the P-layer adjacent to the active layer for minimum capacitance of the upper junction. A simple derivation of

$$C = w_{cl} L \sqrt{\frac{q^2 \varepsilon N_i^2}{2kT N_D}} \sqrt{\frac{\frac{N_A N_D}{N_i^2}}{\ln\left(\frac{N_A N_D}{N_i^2}\right) - \frac{qV_g}{kT}}} \tag{5.41}$$

shows that a minimum capacitance is obtained if

$$\frac{N_A N_D}{N_i^2} = \exp\left(1 + \frac{qV_g}{kT}\right) \tag{5.42}$$

which for InP and a 1.55-μm InGaAsP active layer gives $N_A N_D \approx 10^{28}$ cm^{-3}. For a typical doping level $N_D = 10^{18}$ cm^{-3}, one then finds that the optimum doping level of the P-InP layer is very small and on the order of 10^{10} cm^{-3}. At the same time, a very large resistance R is obtained for such a small doping level, and hence one obtains a very large cutoff frequency imposed by the parasitics. The minimum value for the capacitance is 0.5 pF for a laser with 500-μm width and 500-μm length. This will then also lead to a reduction of the lateral leakage current at high frequencies.

The very small doping level mentioned above corresponds with semi-insulating material, and thus we find that such material should be used in order to receive minimum influence from parasitic elements. In this case, however, the displacement current in the semi-insulating material will dominate the junction capacitance, and it is then useful (both for small capacitance C and large resistance R) to increase the thickness d of the semi-insulating layer.

If a large top electrode and an isolator are used as in Figure 5.11, the distributed capacitance and resistance will be determined by the isolator. The capacitance of such an isolator can be relatively large, while the resistance R will be rather small (unless the upper P-InP layers are weakly doped), resulting in a small cutoff frequency. This can seriously degrade the modulation characteristics.

5.6 MICROWAVE EFFECTS

So far it has been assumed that the current or voltage is uniform in amplitude and phase along the laser stripe. This is not necessarily the case anymore at very high modulation frequencies, since current is usually fed in a small area (Figure 5.16), and the delay time from the feed point to the ends of the stripe can be a significant fraction of the modulation period. This effect has been shown theoretically as well as experimentally to cause a length-dependent roll-off in the modulation response [4].

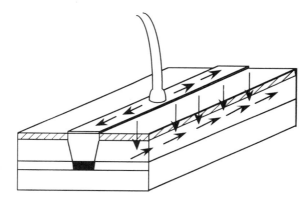

Figure 5.16 Schematic picture of the transverse and longitudinal currents flowing in a laser diode.

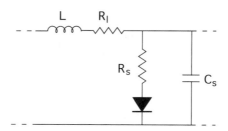

Figure 5.17 Distributed circuit model of the microwave propagation of current and voltage along a laser diode.

Figure 5.16 shows the microwave currents in the device under high-frequency modulation. The transverse currents pump the active region or are leakage currents, and the longitudinal currents are needed to satisfy the boundary conditions imposed on the propagating electromagnetic wave. The roll-off in the modulation response is mainly due to the losses of the longitudinal currents in the cladding layers with their finite conductivity. The device lengths and frequencies at which the losses are important are determined by the propagation velocity. The structure shown in Figure 5.16 is actually a transmission line on a lossy semiconductor, which can exhibit a low wave velocity due to the large internal inductance resulting from the large skin loss. This inductance greatly depends on the conductivity $\sigma_{n\text{-}cl}$ of the N-type cladding. An equivalent distributed circuit that can be used to model the microwave propagation is shown in Figure 5.17. The longitudinal resistance R_l and the inductance L are given by the following expressions:

$$R_l = \frac{1}{w}\left(\sqrt{\frac{\pi f \mu}{\sigma_{contact}}} + \sqrt{\frac{\pi f \mu}{\sigma_{n-cl}}}\right)$$

$$L = 2 + \left(\sqrt{\frac{\mu}{4\pi f \sigma_{contact}}} + \sqrt{\frac{\mu}{4\pi f \sigma_{n-cl}}}\right) \tag{5.43}$$

R_s and C_s are the series resistance and parasitic capacitance discussed in Section 5.5.

5.7 CIRCUIT MODELING OF LEAKAGE AND PARASITICS

As the previous discussions indicate, only a very complex physical model could include an accurate description of all electronic and optoelectronic effects that determine the laser behavior. A valid alternative for such a complex model, which would be based on the solution of the wave equations, the current continuity equations, and Poisson's equation in three dimensions is circuit modeling of both the electronic and optoelectronic behavior. This requires the careful identification of equivalent electronic components such as diodes, transistors, thyristors, resistors, inductances, and capacitances and microwave components such as transmission lines. In general, the parameters of these components will not be easily determined but can for a given fabricated structure be extracted from comparison of measurements and circuit analysis (using commercially available circuit simulators). Such a circuit analysis will also allow one to identify the components that limit the dynamic performance or cause large leakage currents and to modify the design of the laser structure for better performance accordingly.

Current circuit simulation tools do not include optoelectronic components, and consequently an equivalent electronic representation of the optoelectronic processes inside the active layer has to be defined. A simple electronic description can be found if the intrinsic laser behavior is governed by the single-mode rate equations [12]. It is based on the definition of a voltage that represents the carrier density in the active layer and a current that represents the average photon density. A longitudinal description of the wave propagation can be implemented in a circuit simulator by a cascade of transmission lines, even for DFB and DBR lasers and taking into account several longitudinal modes [13].

References

[1] Dutta, N. K., D. P. Wilt, and R. J. Nelson, "Analysis of Leakage Currents in 1.3mm InGaAsP Real-Index-Guided Lasers," *IEEE J. Lightw. Tech.*, Vol. 2, June 1984, pp. 201–208.

[2] Lin, M., S. Wang, and N. Dutta, "Measurements and Modeling of the Harmonic Distortion in InGaAsP Distributed Feedback Lasers," *IEEE J. Quant. Electron.*, Vol. 26, June 1990, pp. 998–1004.

[3] Nagarajan, R., M. Ishikawa, T. Fukushima, R. Geels, and J. Bowers, "High Speed Quantum-Well Lasers and Carrier Transport Effects," *IEEE J. Quant. Electron.*, Vol. 28, October 1992, pp. 1990–2007.

[4] Tauber, D., R. Spickermann, R. Nagarajan, T. Reynolds, A. Holmes, and J. Bowers, "Inherent Bandwidth Limits in Semiconductor Lasers Due to Distributed Microwave Effects," *Appl. Phys. Lett.*, Vol. 64, March 1994, pp. 1610–1612.

[5] Sze, S. M., *Physics of Semiconductor Devices*, 2nd edition, New York: Wiley, 1981.

[6] Landolt, Börnstein, *Semiconductors: Intrinsic Properties of Group IV Elements and III-V, II-VI and I-VII Compounds*, New Series III, Vol. 22a, Berlin: Springer-Verlag, 1982.

[7] Joyce, W., and R. Dixon, "Analytic Approximations for the Fermi Energy of an Ideal Fermi Gas," *Appl. Phys. Lett.*, Vol. 31, September 1977, pp. 354–356.

[8] Casey, H. C., and M. B. Panish, "Heterostructure Lasers," New York: Academic Press, 1978.

[9] Lampert, M. A., and P. Mark, "Current Injection in Solids," New York: Academic Press, 1970.

[10] Agrawal, G. P., and N. K. Dutta, *Long-Wavelength Semiconductor Lasers*, New York: Van Nostrand Reinhold, 1986.

[11] Boyce, W. E., and R. C. DiPrima, *Elementary Differential Equations and Boundary Value Problems*, New York: Wiley, 1977, p. 26.

[12] Tucker, R. S., and I. Kaminov, "High-Frequency Characteristics of Directly Modulated InGaAsP Ridge Waveguide and Buried Heterostructure Lasers," *IEEE J. Lightw. Tech.*, Vol. 2, August 1984, pp. 385–393.

[13] Lowery, A. J., "New Dynamic Model for Multimode Chirp DFB Semiconductor Lasers," *IEE Proc. Pt. J*, Vol. 137, 1990, pp. 293–300.

Chapter 6

The Spectrum of DFB Laser Diodes

The optical spectrum of a laser diode is one of the characteristics that are measured at an early stage after fabrication and that are provided with nearly every single laser diode that is sold or used in an application. Its measurement is rather straightforward and yet provides essential information such as the mode discrimination, the degree of coherence, and the tunability.

A large side-mode rejection is required for almost all lasers used as the transmitter or local oscillator in advanced optical communication systems. As a matter of fact, to guarantee single-mode operation even under modulation, a side-mode suppression ratio (SMSR) (defined as the ratio of the output power emitted in the strongest mode and that emitted in the second strongest mode and expressed in decibels) as high as 30 dB is usually required, and lasers with such high SMSR are called dynamic single-mode. The side-mode rejection depends on the biasing conditions, and theoretically, because the stimulated emission in the main mode increases faster than the amplified spontaneous emission (ASE) in a side mode, a rise of the side-mode rejection with bias level can be expected. This is indeed observed for most Fabry-Perot and DBR lasers, but DFB lasers can exhibit a completely different behavior. DFB lasers that are single-mode at low power levels can become multimode at higher power levels due to longitudinal spatial hole burning (i.e., the axial nonuniformity of the optical power and its influence on the distributed reflections). Obviously, the early measurement of the spectrum allows an early selection of stable dynamic single-mode lasers among all fabricated devices.

A high degree of coherence is also desirable for the lasers applied in spectroscopy and most other measurements, in LIDAR systems, and coherent communications. This requires the width of the main mode (the so-called linewidth) to be small in addition to single-mode behavior. The coherence, which has been one of the main concerns of laser manufacturers for many years, is, however, discussed in much detail in Chapter 9 and will not be treated here. Tunable lasers have an application field that is almost similar to that of coherent lasers. In coherent optical communication, they can be needed as local oscillators for heterodyne detection or as transmitters whose emission wavelength must be tunable to the wavelength of a selected channel. The sensitivity of such systems depends largely on the degree of coherence for both transmitter and local oscillator. In measurements, tunable lasers allow the determination of electro-optic properties of the measured systems as a function of wavelength, while the degree of coherence determines the resolution.

The emphasis of this chapter is on side-mode rejection in general and the design of lasers with stable single-mode behavior in particular. After having described the typical features in the optical spectrum of laser diodes, we will give an in-depth treatment of different factors that can affect the side-mode rejection or the yield of single-mode devices and how they depend on the laser design. The detrimental influence of spatial hole burning has led to the investigation of possible new laser structures with substantially reduced spatial hole burning. A variety of new, special laser structures have been discovered and, therefore, a second major part of this chapter consists of descriptions of these special laser structures. This is followed by a brief introduction to the tunability. Different types of tunable lasers have been described extensively in the literature, but our discussion is limited to multielectrode DFB lasers and we refer to the references for more details on this topic. We conclude by briefly describing the measurement of the spectrum and show how a spectrum, measured at a current below the threshold current, can be used to extract the parameters of fabricated devices after fitting an analytical expression to it.

6.1 AMPLIFIED SPONTANEOUS EMISSION

As described in Chapter 1, a laser diode can be conceived as an optical amplifier with positive feedback (provided by reflection at the facets or distributed reflection), with the spontaneous emission as input signal and amplified and filtered spontaneous emission as the (optical) output signal. In fact, the presence of spontaneous emission always prevents the amplification or stimulated emission from compensating for the total loss (including, for example, the power emitted throught the facets) and thus prevents the amplifier from becoming a true oscillator (with self-sustained oscillations). If this were not the case, the input of spontaneous emission would give rise to an infinite number of photons and an infinite stimulated emission rate for which the finite injected current could not supply sufficient carriers. The amplification, however, will approach the loss more and more closely as the bias applied to the laser diode is increased.

The amplification and filtering processes are easily described analytically for Fabry-Perot lasers. A one-dimensional model illustrating the different processes is given in Figure 6.1. Assuming that the spontaneous emission rate R_{sp} is uniformly

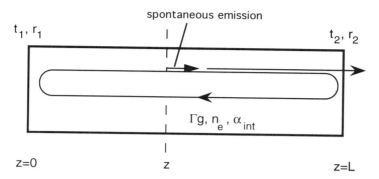

Figure 6.1 One-dimensional model for spontaneous emission and propagation in a single-section Fabry-Perot cavity.

distributed along the laser length L and between the positive and negative directions, one finds a spontaneous emission rate $R_{sp}/2L$ per unit distance and per direction. To calculate the resulting intensity at the facet on the right-hand side, we further assume a uniform gain Γg and internal loss α_{int} per unit distance, a uniform refractive index n_e, and field reflectivities r_1 and r_2 at the facets. The field of a spontaneous emission at a point z and in the + direction then first propagates toward facet 2. At this facet, with coordinate $z = L$, the field is given by

$$E_1(L) = t_2 E_{sp}(z) \exp\left\{\frac{\Gamma g - \alpha_{int}}{2}(L-z)\right\} \exp\{-jkn_e(L-z)\}, \quad k = \frac{2\pi}{\lambda} \quad (6.1)$$

where it is partly transmitted (with transmission t_2) and partly reflected (with reflection r_2). The reflected part then propagates further toward facet 1, is partly reflected (with reflection r_1), and propagates back to facet 2, where again the whole cycle is repeated. The field after n round trips inside the cavity is therefore

$$E_n(L) = E_1(L)\{r_1 r_2 \exp[(\Gamma g - \alpha_{int})L] \exp[-2jkn_e L]\}^n \quad (6.2)$$

and for the total field at facet 2 caused by the spontaneous emission at a point z in the + direction, one finds

$$E(L) = E_1(L) + E_2(L) + E_3(L) + \ldots = \frac{E_1(L)}{1 - r_1 r_2 \exp[(\Gamma g - \alpha_{int})L] \exp[-2jkn_e L]} \quad (6.3)$$

Obviously, a spontaneous emission in the − direction also contributes to the field at facet 2. This contribution can be calculated in a similar way. It is not correlated with the field caused by the emission in the + direction and thus the total intensity is the sum of the intensities of the two contributions. The total intensity caused by all spontaneous emissions is just the sum of the intensities resulting from each single spontaneous emission and is given by (with $R_2 = |r_2|^2$)

$$P_{tot}(L) \sim \int_0^L dz \, \frac{\exp[(\Gamma g - \alpha_{int})(L - z)] + R_2 \exp[(\Gamma g - \alpha_{int})(L + z)]}{|1 - r_1 r_2 \exp[(\Gamma g - \alpha_{int})L] \exp[-2jkn_e L]|^2} \quad (6.4)$$

From this expression, it follows that the intensity is maximum at the mode wavelengths (i.e., if $m\lambda = 2n_e L$) and minimum for $(m + 1/2)\lambda = 2n_e L$. It also follows that the gain approaches the total loss as the power increases; that is,

$$\Gamma g \to \alpha_{int} + \frac{1}{L} \ln\left(\frac{1}{r_1 r_2}\right) \text{ as } P_{tot} \to \infty$$

It can, of course, be seen that we have ignored the dependence of gain and refractive index on the electron density and the dependence of the electron density on the optical power density in our derivation. These dependencies cause the gain and refractive index to be nonuniform and, more importantly, lead to a power dependence of the phase factor in the denominator that results in a broadening of the resonance peaks. The expression (6.4) is therefore only very accurate as long as the laser diode is biased sufficiently below threshold. Close to or above threshold it can be considered as an implicit equation for the spectral dependence of the optical power if the exact dependence of gain and refractive index on the optical power is taken into account.

The calculated optical spectrum of a typical Fabry-Perot laser biased below threshold is shown in Figure 6.2. The round-trip gain of this laser calculated at threshold is shown in Figure 6.3. The amplitude of the round-trip gain is $r_1 r_2 \exp[(\Gamma g - \alpha_{int})L]$ and gives an indication of the dispersion in the gain. The parameters used in the calculation are given in Table 6.1. The mirror losses in Fabry-Perot lasers are equal for all modes and hence the mode discrimination can have its origin only in the wavelength dependence of the gain and the internal loss. The internal loss, consisting of free carrier and intervalence band absorption and of scattering, depends little on the wavelength in practice and has been assumed constant in the calculations. The dispersion in the gain then remains as the only mode discriminating factor for Fabry-Perot lasers.

From (6.4), it can be predicted that the output power of the side modes is nearly constant above threshold and that the SMSR increases nearly proportionally to the

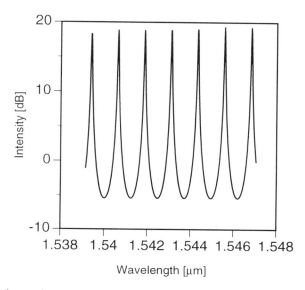

Figure 6.2 Calculated optical spectrum at threshold for the Fabry-Perot laser with parameters given in Table 6.1.

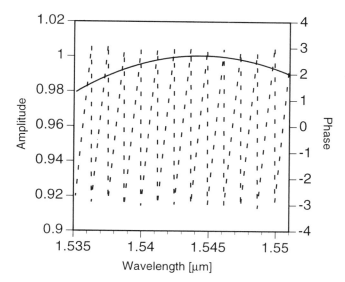

Figure 6.3 Amplitude (solid line) and phase (dashed line) of the round-trip gain for a 300-μm-long cleaved Fabry-Perot laser.

Table 6.1
Typical Parameter Values for a Long-Wavelength Laser Diode

Parameter	(Unit)	Typical Value	Parameter
w	(µm)	1.5	Stripe width
d	(µm)	0.12	Thickness active layer
Γ		0.5	Confinement factor
L	(µm)	300	Laser length
Λ	(µm)	0.2413	Grating period
n_e		3.25	Effective refractive index of the unperturbed waveguide
η		0.8	Injection efficiency
τ	(sec)	5×10^{-9}	Carrier lifetime
B_0	(µm³/s)	100	Bimolecular recombination
C_0	(µm⁶/s)	20×10^{-5}	Auger recombination
β_{sp}		10^{-4}	Spontaneous emission factor
α_{int}	(µm⁻¹)	50×10^{-4}	Internal absorption loss

output power in the main mode. Above threshold, the gain of the main mode is nearly equal to the total loss, and the gain (depending on the electron density) does not change very much more with power level. The gain changes for the side modes are equally small and much smaller than the gain change that would be required to compensate for the loss. Hence, the round-trip gain for the side modes remains almost constant. Although one must bear in mind that, due to the gain suppression, the dispersion in the gain may change with bias level, it can be argued that gain suppression (typically on the order of 10^{-3}/mW) only has a minor effect on the static behavior. Spatial hole burning has an identical effect on all Fabry-Perot modes and does not change the mode rejection either.

The difference in gain between the main mode and a side mode is largely determined by their wavelength difference. Indeed, the curvature of the $g(\lambda)$ curve is, for a given material, practically independent of the carrier density and thus of the threshold gain. The side-mode suppression can thus be improved by an increase in the mode separation ($\Delta\lambda = \lambda^2/2n_e L$), that is, by decreasing the laser length.

The output power in main and side modes as a function of the injected current for the considered Fabry-Perot laser is shown in Figure 6.4. The figure confirms that the power in the side modes remains constant above threshold. It also indicates that the SMSR in Fabry-Perot lasers of reasonable length is usually well below 20 dB.

The calculation of the spectrum of DFB lasers can be done in a similar way, although the presence of distributed reflections makes it less straightforward. The

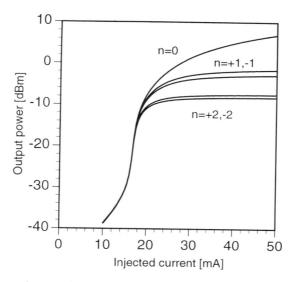

Figure 6.4 Output power of main and side mode vs. injected current for a 300-μm-long cleaved Fabry-Perot laser.

propagation toward facet 2 of a spontaneously emitted wave E_{sp} at z in the + direction can now be calculated with the F-matrix or transfer matrix (defined in Chapter 3)

$$\begin{pmatrix} E^+(L) \\ E^-(L) \end{pmatrix} = \begin{pmatrix} F_{11}(L-z) & F_{12}(L-z) \\ F_{21}(L-z) & F_{22}(L-z) \end{pmatrix} \begin{pmatrix} E^+(z) \\ E^-(z) \end{pmatrix} \tag{6.5}$$

This transfer matrix F (which is not the same as the one used in Chapter 4, as it defines the relationship between the fields and not between their complex amplitudes $R^{\pm}$) has a determinant equal to 1 and is obviously transitive (i.e., $F(L-z)F(z) = F(L)$). Elimination of the backward propagating fields from these equations (using the boundary condition $E^+(L) = r_2 E^-(L)$) gives the field $E_1(L)$ resulting from the first propagation of $E_{sp}(z)$ as

$$E_1^+(L) = t_2 E_{sp}(z) \left(\frac{\det(F)}{F_{22} - r_2 F_{12}} \right)_{L-z} = t_2 E_{sp}(z) \frac{1}{F_{22}(L-z) - r_2 F_{12}(L-z)} \tag{6.6}$$

and the total field at $z = L$ caused by this spontaneous emission is obtained after division of the last expression by 1 minus the round-trip gain. The contribution of a spontaneous emission at z in the $-$ direction can be calculated with the transfer matrix $F(z)$ (instead of $F(L-z)$). Use of the boundary condition at the left facet gives the forward propagating field $E^+(z)$ corresponding with $E^-(z)$ ($= E_{sp}(z)$), and the first propagation of $E^+(z)$ toward the right facet results in

$$E_1^-(L) = t_2 E_{sp}(z) \left(\frac{r_1 F_{11}(z) + F_{12}(z)}{r_1 F_{21}(z) + F_{22}(z)} \right) \frac{1}{F_{22}(L-z) - r_2 F_{12}(L-z)} \tag{6.7}$$

The total field at $z = L$ is again obtained after division by 1 minus the round-trip gain. The round-trip gain in a DFB laser depends in general on the position z where it is considered. An expression for it is given by the product of the reflectivities r^+ and r^- experienced from the left and right sides:

$$r^+ r^- = - \frac{F_{21}(L-z) - r_2 F_{11}(L-z)}{F_{22}(L-z) - r_2 F_{12}(L-z)} \frac{F_{11}(z) r_1 + F_{12}(z)}{F_{21}(z) r_1 + F_{22}(z)} \tag{6.8}$$

The total intensity emitted through the facet on the right-hand side and resulting from all spontaneous emissions can then again be calculated by adding the intensities resulting from each separate spontaneous emission. After some simple calculations, one then finds for the total intensity at the right facet:

$$P_{tot}(L) \sim \int_0^L dz \, \frac{|F_{21}(z) r_1 + F_{22}(z)|^2 + |F_{11}(z) r_1 + F_{12}(z)|^2}{|F_{22}(L) + r_1 F_{21}(L) - r_2 F_{12}(L) - r_1 r_2 F_{11}(L)|^2} \tag{6.9}$$

Again, the wavelength dependence of the emitted power is mainly due to the wavelength dependence of the round-trip gain (discussed in Section 4.1). The nature of

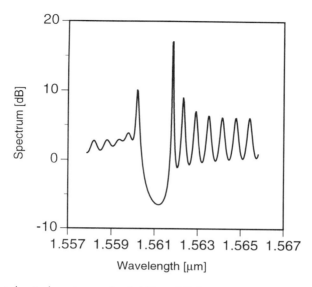

Figure 6.5 Calculated optical spectrum at threshold for a DFB laser.

distributed feedback now implies a much stronger wavelength dependence of the round-trip gain and therefore also a much better side-mode rejection. The spectrum of a typical DFB laser biased at threshold calculated using (6.9) is shown in Figure 6.5. A laser length of 300 μm, a κL value of 1, and facet field reflectivities equal to $r_1 = 0.566\ e^{j\pi}$ and $r_2 = 0.224\ e^{j3\pi/2}$ were assumed for this laser. The calculated round-trip gain at threshold of this laser can be seen in Figure 6.6.

6.2 SIDE-MODE REJECTION AND YIELD OF DFB LASERS

An estimate of the side-mode suppression in DFB lasers at low power levels can be obtained from the threshold gain difference ΔgL, that is, the difference in gain required for compensation for the loss between the main mode and the first side mode. The value of ΔgL greatly depends on the laser structure under consideration and on the normalized coupling constant κL and varies from 0 for AR-coated, index-coupled lasers to more than 1 for some λ/4-shifted or gain-coupled lasers. As outlined in Chapter 2, a ΔgL value of 0.1 or more would (if spatial hole burning were absent) guarantee single-mode operation (with an SMSR of 30 dB) at ± 1 mW of output power.

In DFB lasers with partly reflecting facets, reflections from the grating and

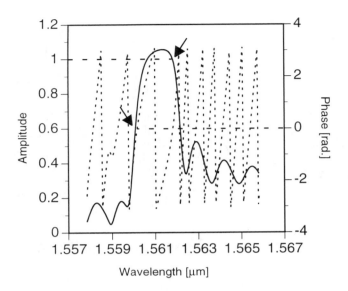

Figure 6.6 Calculated round-trip gain at threshold for the DFB laser considered in Figure 6.5.

facets will interfere. The interference depends on the phase of the grating at the facets and has a significant influence on the wavelength dependence of both the amplitude and the phase of the round-trip gain. Depending on the phase of the grating at both facets, one can obtain either single-mode or multimode operation. For example, if, in an index-coupled laser, the grating phases at both facets result in a symmetric structure, a mode degeneracy similar to that in the AR-coated laser exists because of the symmetry. For some other grating phases, however, the interference lifts the degeneracy, and single-mode behavior is obtained. The interference between grating reflections and facet reflections causes the threshold gain difference ΔgL to depend on the phase of the grating at the facets.

The current technology (i.e., the cleaving or etching processes) does not yet allow the control of these phases, and one can say that the phases are distributed randomly among the different lasers grown on one wafer (Figure 6.7). This random distribution of phases gives rise to a random distribution of ΔgL values and a single-mode yield (the percentage of lasers with a ΔgL value larger than a certain predefined value) is therefore defined.

The Spectrum of DFB Laser Diodes 157

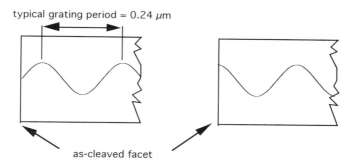

Figure 6.7 Statistical phase between grating and facet for two devices from the same wafer.

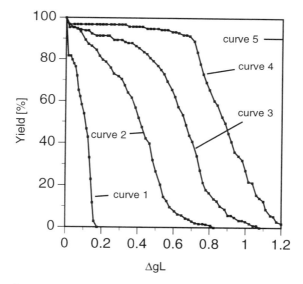

Figure 6.8 Single-mode yield of partly gain-coupled DFB lasers, all for $|\kappa|L = 2$. Curve 1: index coupling with 32% reflection; curve 2: $\kappa_g/\kappa_n = 0.58$ with 32% reflection; curve 3: $\kappa_g/\kappa_n = 0.58$ with 10% reflection; curve 4: $\kappa_g/\kappa_n = 0.58$ with 4% reflection; curve 5: $\kappa_g/\kappa_n = 0.58$ with 0% reflection. (*After:* [4], © 1991 IEEE.)

The yield has already been studied in much detail (e.g., see [1–3]). The calculated numbers for the yield are shown as a function of ΔgL in Figure 6.8 for different values of facet reflectivity, index coupling, and gain coupling. Gain-coupled lasers obviously show much larger yields than index-coupled lasers. This larger yield can be attributed to the presence of a standing-wave pattern, both in the power of the cavity modes (with period $\lambda/2n_e$) and in the gain (period Λ). The overlap of both standing-wave patterns increases for a decreasing difference between mode wavelength and Bragg wavelength. The influence of this overlap can easily be seen from the rate equations as derived in Chapter 4. Apart from the cavity loss γ, the standing wave term $2\kappa_g f_{st}$ now also depends on the wavelength or the mode number m in the rate equation for the photon density.

A comparison between the experimentally obtained yield (i.e., the percentage of single-mode lasers) and the calculated yield (requiring $\Delta gL > 0.05$) as a function of κL for as-cleaved lasers was discussed in [4] and is shown in Figure 6.9 for index-coupled lasers. The experimental yield seems to be optimum for a κL value around 1 and decreases again for larger κL values, whereas the yield based on the ΔgL number increases monotonically with κL. The deviation between theory and experiment indicates that another factor, spatial hole burning, must be taken into account.

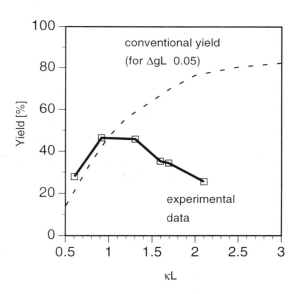

Figure 6.9 Experimental (solid line) and calculated (dashed line) yield of as-cleaved DFB lasers vs. κL. (*After:* [4], © 1991 IEEE.)

6.3 DEGRADATION OF THE SMSR BY SPATIAL HOLE BURNING

A DFB laser with a relatively high ΔgL value and being single-mode at low power levels can become multimode at moderate or high power levels due to spatial hole burning [5]. As an example, we consider again a 600-μm-long laser with a κL value of 3 and with the field reflectivities of the facets being given by $r_1 = 0.566\ e^{j\pi}$ and $r_2 = 0.224\ e^{j3\pi/2}$. The parameters are identical to those of Table 6.1. This laser has a threshold current of 39.6 mA and a normalized threshold gain difference between main and first side modes of 0.25. But in spite of this large ΔgL value, one finds that the side mode reaches threshold at about 70 mA, corresponding with an ex-facet power of 2 mW (Figure 6.10). This side-mode onset is also illustrated by the wavelength dependence of the round-trip gain calculated at 70 mA, which is displayed in Figure 6.11.

Evidence for spatial hole burning (i.e., the variation of the carrier density due to the variation of the power) being the cause of the side-mode onset can be given as follows. The z-dependence of the carrier density causes the effective refractive index

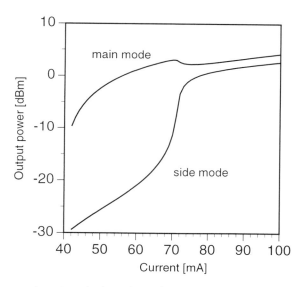

Figure 6.10 Output power in main and side modes vs. the injected current for a DFB laser with a side-mode onset.

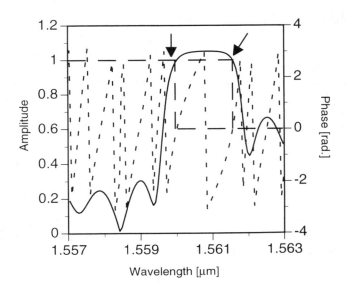

Figure 6.11 Amplitude (solid line) and phase (dashed line) of the round-trip gain of the laser of Figure 6.9 at 70 mA.

n_e, the Bragg wavelength ($\lambda_B = 2n_e\Lambda$), and the real part of the Bragg deviation $\Delta\beta_r$ also to become z-dependent. The reflection losses, however, depend on $\Delta\beta_r$ and, since $\Delta\beta_r$ assumes various values along the z-axis, the losses become less dependent on the wavelength. Or, in other words, a broader range of wavelengths are undergoing strong distributed reflections within a smaller fraction of the entire grating as the power increases. For this specific laser, the power of the main mode is concentrated in the center of the laser. Hence, the carrier depletion in the center is considerably larger than the average carrier depletion. Since the average carrier density remains more or less constant as the power increases, a relatively large carrier density near the facets develops with increasing power. This corresponds with a locally smaller Bragg wavelength, which, at sufficiently high excitation, coincides with the wavelength of the side mode. The side mode is then strongly reflected and experiences noticeably lower losses than at threshold.

$\Delta\beta_r$ is depicted in Figure 6.12 for both the main and side modes and for different injection levels. One can see how the average Bragg deviation increases for the main mode, while it decreases for the side mode. The first phenomenon is accompanied by an increase in the loss of the main mode and thus, due to gain clamping, an increase in the average modal gain for both the main and side modes. The second phenomenon results in a decreased loss of the side mode, which eventually makes it reach the threshold.

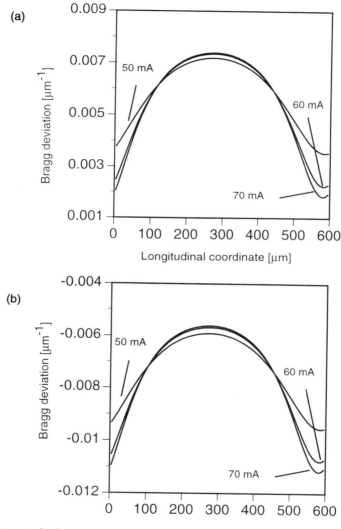

Figure 6.12 Longitudinal variation of the Bragg deviation of the laser of Figure 6.9 at different injection levels: (a) main mode and (b) side mode.

The impact of spatial hole burning on the side-mode onset depends both on the nonuniformity of the carrier density and on the degree to which this nonuniformity is reflected in a nonuniformity of the refractive index and the Bragg deviation. From (3.4) for the nonuniform carrier density,

$$\Delta N(z) = - \frac{G_{\text{th}} \Delta S(z)}{\dfrac{1}{\tau_d} + \dfrac{\partial G}{\partial N} S_0} \tag{6.10}$$

with S_0 the average photon density, $\Delta S(z)$ the axial variation in the photon density, and $G = \Gamma g v_g$, it readily becomes clear which parameters are of importance:

- The nonuniformity of the optical power (expressed by ΔS);
- The power level (expressed by S_0);
- The losses (expressed by the average gain G_{th});
- The differential gain $\partial G/\partial N$ and the differential carrier lifetime τ_d;
- The *differential* refractive index $\partial n/\partial N$;
- The wavelength separation between main and side modes.

The nonuniformity of the power depends on the specific laser structure and on the κL value. Lasers with high κL value tend to become rapidly multimode due to the strongly nonuniform power. Indeed, the distributed reflections are stronger for higher κL values, and, in general, the power seems to be concentrated in the central region of the laser. In such lasers it is often the side mode on the short wavelength side of the Bragg wavelength that can reach the threshold. This mode has a positive Bragg deviation at threshold, while the concentration of the power in the center of the laser corresponds to a concentration of the electron density near the facets. Since the refractive index decreases with increasing carrier density, it follows that the Bragg deviation of each mode decreases near the facets and that the Bragg deviation of the side mode on the short wavelength side approaches zero. A schematic representation of this process is given in Figure 6.13.

It must, however, be noted that higher κL values also imply smaller facet losses and thus a relatively small nonuniform carrier density. The loss can further be reduced by decreasing the internal absorption or scattering. The nonuniformity of the power is not affected in this case. The absorption depends not only on the quality of the grown active and passive layers, but also on the carrier density (which itself depends on the loss and gain parameters) and on the wavelength [6].

The nonuniform carrier density can also be reduced by reducing the carrier lifetime or, to a lesser extent, by increasing the differential gain [7]. Both quantities

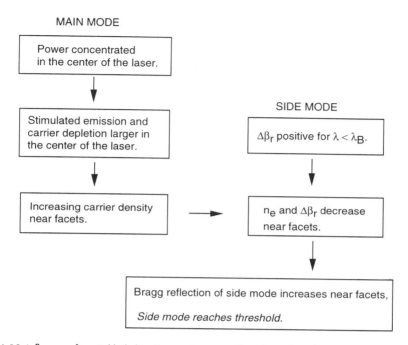

Figure 6.13 Influence of spatial hole burning on the onset of a side mode with $\lambda < \lambda_B$ for large κL.

depend on the chosen material (e.g., bulk or quantum-well material), the temperature, and the doping levels in the active layer. The differential gain in quantum wells (and especially in strained quantum wells), furthermore, depends on the threshold gain (or more generally, on the laser structure) and increases with decreasing threshold carrier density because of the sublinear dependence of gain on carrier density.

A change in the carrier dependence of the refractive index, on the other hand, reduces the nonuniform index or Bragg deviation instead of the nonuniform carrier density. As shown before, it is the nonuniformity in the Bragg deviation that causes the side-mode onset. The carrier density dependence of the refractive index again depends on the semiconductor material, the temperature, and the doping levels.

One can see that the differential gain in (6.10) includes the confinement factor and will therefore increase proportionally to the confinement factor. Increasing the confinement factor will, however, also cause a proportional increase of the carrier density dependence of the effective refractive index, and hence it seems more beneficial to decrease the confinement factor.

Finally, it must be noticed that shorter lasers are more stable with respect to the side-mode onset. In many cases, the Bragg deviation of the side mode is inversely proportional to the laser length and thus larger for shorter lasers. A larger nonuniformity of the refractive index is then required to cause a side-mode onset.

All of the dependencies reported here have indeed been confirmed, by experiments as well as by our simulations. As an illustration we show in Figure 6.14 the output power in main and side modes for a laser, that is identical to that of Figure 6.10 except that it is only half as long. One finds stable single-mode behavior up to very high power levels, and the side-mode suppression does not drop below 20 dB, even at an output power of 20 mW.

According to (6.10), spatial hole burning saturates at high power levels. In the derivation of (6.10), it was, however, assumed that the relative longitudinal variation of the optical power in the main mode remains virtually unchanged above threshold. This is not always the case and the longitudinal mode profile of the main mode may itself be unstable. This aspect is discussed in detail in [8], and we will only briefly review it here.

We consider a 300-µm-long DFB laser with $\kappa L = 2$ and facet reflectivities $r_i = -j0.566$ and $r_2 = 0$. Figure 6.15 shows the longitudinal variation of the power in the main mode at threshold and at 5-mW bias output power. One can see that the power in the main mode becomes concentrated more and more near the facet at $z = L$ as the power increases. This is easily understood from the threshold characteristics. At threshold, the Bragg deviation of the laser is positive and the power is concentrated

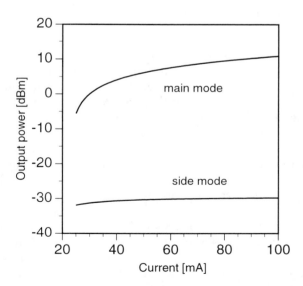

Figure 6.14 Output power in main and side modes vs. injected current for a DFB laser with stable side-mode rejection.

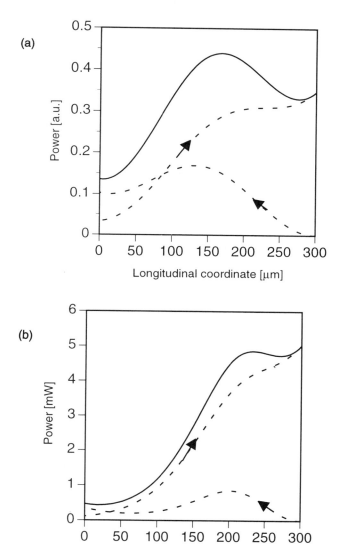

Figure 6.15 Axial power variation in a DFB laser with unstable power profile: (a) at threshold and (b) at 5 mW.

near $z = L$. Hence, above threshold, the carrier density will be depleted near $z = L$ and an increased Bragg deviation will result. This in turn implies that the Bragg reflections are weakened near $z = L$ and intensified near $z = 0$., so that the power becomes more concentrated near $z = L$.

This effect occurs mainly in lasers with strongly asymmetric facets or when the main mode emits at the short (long) wavelength side of the Bragg wavelength for lasers with a high (low) κL value. Obviously, this relatively strong spatial hole burning easily causes the onset of side modes.

Spatial hole burning effects can be taken into account in yield calculations by imposing an extra condition related to the uniformity of the optical power. In [4], a much better agreement between experimental and theoretical yield was found by requiring a certain minimum value for an f-number, defined as the ratio of the minimum to the maximum power density along the longitudinal direction. Figure 6.16 [4] shows the theoretical yield obtained for different minimum values of the f-number for index-coupled lasers with two cleaved facets. A ΔgL value of 0.05 was required. Comparison with the experimental yield in Figure 6.9 now gives a much better agreement. Figure 6.17 gives the results for partly gain-coupled lasers with a fixed gain coupling of 26.18 cm^{-1}. In this case, a ΔgL value as high as 0.3 was required and much higher f-numbers are considered. These high values for f and for ΔgL again reflect the high single-mode yields that can be obtained for partly gain-coupled lasers.

A final remark about the impact of spatial hole burning concerns its dependence on the series resistance of the laser. So far we have assumed that the injected current

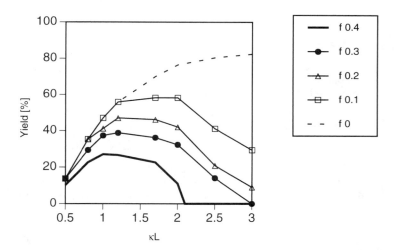

Figure 6.16 Spatial hole burning–corrected yield for cleaved lasers as a function of κL and for $\Delta gL > 0.05$. (*After:* [4], © 1991 IEEE.)

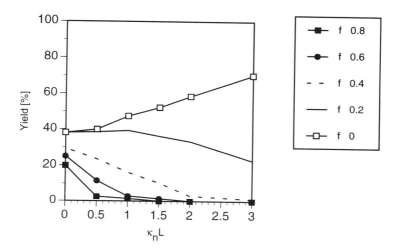

Figure 6.17 Spatial hole burning–corrected yield for cleaved lasers as a function of $\kappa_n L$ with $\kappa_g = 26.18$ cm^{-1} and for $\Delta gL > 0.3$. (*After:* [4], © 1991 IEEE.)

density is uniform along the laser axis, while in reality this current density depends on the voltage over the heterojunction and the resistivity of the cladding layers (as discussed in Chapter 4) [9]. The voltage over the double heterojunction will in fact be large and allow less current where the carrier density is large and be small and allow more current where the carrier density is small. This effect, which counteracts spatial hole burning, is obviously more pronounced if the series resistance is lower. It is also an easy matter to include it in (6.10) for the nonuniform carrier density $\Delta N(z)$. The inverse differential carrier lifetime τ_d must simply be replaced by

$$\frac{1}{\tau_d} + \frac{1}{qR_s wdL} \frac{dV_{\text{DH}}(N_0)}{dN} \qquad (6.11)$$

in which R_s is the series resistance and V_{DH} the voltage over the double heterojunction.

6.4 DFB LASERS WITH REDUCED SPATIAL HOLE BURNING

A relatively high yield of stable single-mode devices can result after fabrication of rather short DFB lasers with partly reflecting facets and a "small" κL value. And a stable single-mode behavior is even guaranteed for all AR-coated, λ/4-shifted lasers with a rather small κL value and short length. However, one sometimes prefers longer AR-coated DFB lasers for some applications (e.g., if a small linewidth and a very high yield are required). To avoid the onset of side modes due to spatial hole burning in such lasers, several special structures and methods have been and are still being investigated. It would lead us too far from the subject to describe them all in detail, but a selection is given below. The reader is referred to the specialized literature for more detail on some of the structures or methods.

6.4.1 Nonuniform Injection

This obviously requires the use of multielectrode lasers, where the injection into each electrode is adjusted until single-mode behavior is achieved. Numerous examples of this approach have already been given in the literature [10–12]. We repeat that such laser types are also most suitable for tuning purposes.

As an example, we consider a λ/4-shifted laser with $L = 300$ µm, $\kappa L = 3$, and other parameters as given in Table 6.1. The stripe electrode has been divided into three parts, as shown in Figure 6.18, although the first and third parts are short-circuited. The laser becomes multimode at ±80 mA when uniformly pumped ($I_c = I_e$ in Figure 6.19). But one easily finds that this can be resolved by pumping the central electrode more strongly than the outer electrodes, which is illustrated in Figure 6.19 for the case $I_c = 1.2 I_e$. Since the optical power is concentrated in the central region of this laser, there is also a larger stimulated emission and hence a larger carrier depletion in this region. Pumping relatively more carriers in this central region, however,

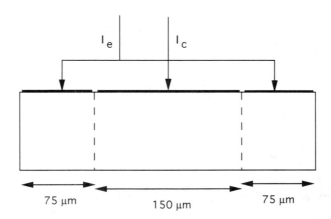

Figure 6.18 Schematic view of a multielectrode laser.

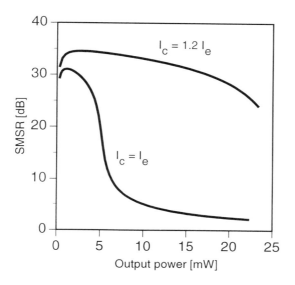

Figure 6.19 Power in main and side modes for uniform ($I_c = I_e$) and nonuniform ($I_c = 1.2 I_e$) injection for the laser of Figure 6.18.

prevents the carrier density from becoming too nonuniform and therefore the laser from becoming multimode.

6.4.2 Special Index-Coupled Structures

Many of the proposed special index-coupled structures, which aim at a more uniform power distribution along the longitudinal axis, are multiphase-shifted lasers (i.e., with several phase shifts in the grating) and chirped grating lasers (with gratings with a varying period). A drawback of these lasers is that in general they require the grating to be written by an e-beam and hence are not so easily fabricated. Multiphase-shifted lasers were originally proposed by Agrawal et al. [13], who proved that two or three properly chosen phase shifts at equispaced points along the axis give a far more uniform power than one phase shift of λ/4 and still have a sufficient threshold gain difference. This analysis was later extended by several groups to cases with nonequispaced and nonidentical phase shifts (see [14,15]). The fabrication of such devices has been reported in, for example, [15,16]. A stable single-mode behavior (and usually a very narrow linewidth) was found even for long cavity lengths (e.g., 1,200 μm or more). As an example, we show the longitudinal variation of the power for a multiphase-shifted laser in Figure 6.20. The laser under consideration has a length of

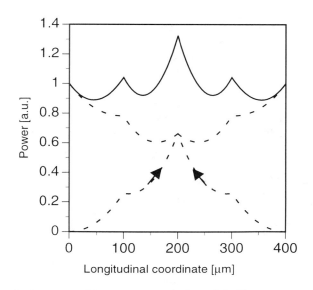

Figure 6.20 Longitudinal variation of the power in a multiphase-shifted laser.

1,200 µm and a normalized coupling constant of 2. Three phase shifts of $4\pi/5$ have been incorporated in the grating every 300 µm. The total optical power density along the length varies only ±30%, which, in comparison with common DFB lasers, is an appreciable reduction.

A similar uniformity of optical power can be obtained in chirped-grating- or corrugation-pitch-modulated lasers. The first chirped-grating lasers were so-called phase-adjusted lasers, and they were proposed as early as 1984 by Tada et al. [17]. The strong spatial hole burning of conventional $\lambda/4$-shifted lasers was reduced in these phase-adjusted lasers by replacing the discontinuity in the grating by a built-in steplike nonuniform stripe width. The larger stripe width near the center of the laser results in a locally higher effective refractive index and hence a locally higher Bragg wavelength. Fabricated devices [18,19] of this type exhibited an improved single-mode stability compared with the $\lambda/4$-shifted devices. Real chirped gratings (i.e., with a variable grating period) have been reported since 1990 (e.g., see [20]). Calculations for linearly chirped gratings indicate that spatial hole burning is greatly reduced with increased chirp factor, but that at the same time the threshold gain difference is also severely decreased [21]. The corrugation-pitch-modulated laser described in [20] has been used to demonstrate extremely small linewidths. The structure is depicted in Figure 6.21. An economically interesting method of realizing DFB lasers with arbitrarily chirped gratings was recently proposed [22,23]. It is based on the use of a geometrically uniform grating and the creation of an effective chirp by slightly bending the optical waveguide as illustrated in Figure 6.22. The bend angles must be kept

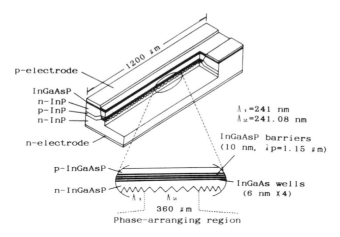

Figure 6.21 Corrugation-pitch-modulated DFB laser. (*Source:* [20], © 1990 IEEE.)

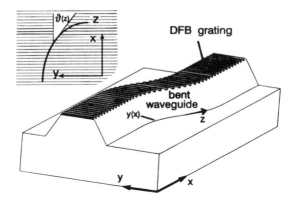

Figure 6.22 Fabrication of chirped gratings using bent waveguides. (*Source:* [22], © 1993 IEEE.)

sufficiently small to limit the additional loss, but this does not prevent the creation of chirps equivalent to a $\lambda/4$ phase shift.

It has been shown in [24,25] that complete elimination of spatial hole burning is possible if an axially varying coupling constant and/or a variable gain/loss are introduced. In fact, structures with perfectly uniform optical power density can be derived easily from the coupled-wave equations, and a variety of functions for the coupling constant κ, the Bragg deviation $\Delta\beta_r$, and the modal gain $\Delta\beta_i$ that give uniform power have already been derived from these coupled-wave equations. Approximating the modal gain or coupling constant variation of some of these solutions with a constant results in pure chirped-grating lasers. A solution that has also been realized is

$$\kappa(z) = \frac{1}{2L} \frac{\left(1 - 2\frac{z}{L}\right)}{\sqrt{\frac{z}{L}\left(1 - \frac{z}{L}\right)}}$$

$$\Delta\beta_r(z) = \frac{1}{4L} \frac{\left(1 - 2\frac{z}{L}\right)}{\frac{z}{L}\left(1 - \frac{z}{L}\right)} c \tag{6.12}$$

with c an arbitrary real constant. For $c = 0$, one finds a laser with a grating whose amplitude varies in the longitudinal direction [25]. The $\kappa(z)$ variation can then, for example, be approximated by a linear, a cosine, or a stepwise constant function. As a matter of fact, the function can be approximated to any degree if gratings are written by e-beam lithography. One can either vary the actual grating amplitude or the duty cycle of the grating, as shown in Figure 6.23. A stepwise constant approximation, together with the exact function, is shown in Figure 6.24 for a 300-μm-long laser. This approximation already results in an extremely uniform optical power (Figure 6.25), with variations that are restricted to 5%. For the threshold gain difference ΔgL, one finds the value 0.17, and therefore stable single-mode behavior should be observed for this structure.

A second approximation can be formed by the double exposure of a photoresist to form two holographic interference patterns of slightly different periods Λ_1 and Λ_2 [26,27]. This results in a cosine variation of κ. However, the variation of the coupling coefficient will in general be accompanied by a variation of the effective refractive index (and of the Bragg deviation) if the last method is used. The relationship between the variation of the coupling coefficient and that of the refractive index depends on the lithography and etching process. One possible structure is shown in Figure 6.26, in which case $\Delta\beta_r = 0.5|\kappa|$. This is fundamentally different from (6.12), which could be approximated somehow as $\Delta\beta_r = c\kappa$.

A last approximation that is worthwhile mentioning is a stepwise constant approximation with κ being constant in the outer sections and zero (i.e., no grating) in the central section. One must thereby ensure that the π-phase shift between both gratings near the facets is still present. One possibility is to fabricate a $\lambda/4$-shifted grating of which a central part is removed afterwards.

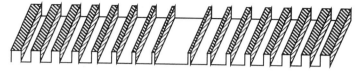

Figure 6.23 Rectangular grating with variable duty cycle.

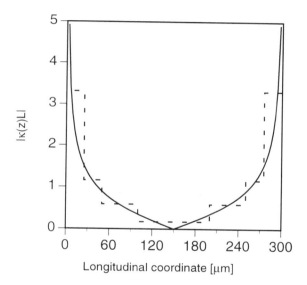

Figure 6.24 Longitudinal variation of κ (6.12) and stepwise constant approximation for $c = 1$.

6.4.3 Gain-Coupled Lasers

The introduction of gain coupling (i.e., a periodic variation in gain or loss) is an alternative method to obtain an axially uniform optical power density in AR-coated lasers [28]. In fact, perfectly uniform power can be obtained in lasers with pure gain coupling. However, relatively uniform power can also be obtained in index-coupled lasers with a small degree of gain coupling. Such lasers are more easily fabricated than purely gain-coupled lasers and therefore of more practical importance [29–31].

The coupled-wave equations for DFB lasers with both index and gain coupling can be written as

174 HANDBOOK OF DISTRIBUTED FEEDBACK LASER DIODES

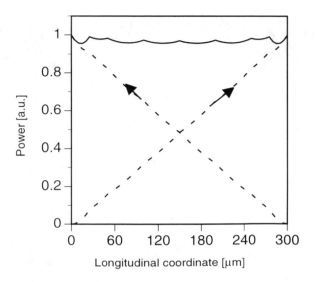

Figure 6.25 Longitudinal variation of the power for the stepwise constant approximation of Figure 6.24.

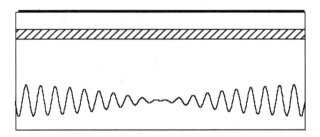

Figure 6.26 Grating formed by the double exposure technique.

$$\frac{dR^+}{dz} + j\Delta\beta\, R^+ = (\kappa_g + j\kappa_n)R^-$$

$$\frac{dR^-}{dz} - j\Delta\beta\, R^- = -(\kappa_g + j\kappa_n)R^+ \qquad (6.13)$$

in which the coupling constants κ_g and κ_n stand for gain and index coupling, respectively. The phases of κ_g and κ_n are assumed to be equal here, which is justified if the gain and index coupling are induced by one grating (e.g., a modulation of the active layer thickness). The phase difference between κ_g and κ_n can then only assume the values 0 and π, whereas a phase difference of π would only cause a change in the sign of the Bragg deviation of the lasing mode. In addition, the grating phase at the left facet is chosen such that κ_g and κ_n are both real.

As a measure of spatial hole burning, we use the ratio P_{min}/P_{max} with $P_{min(max)}$ denoting the minimum (maximum) value of the power along the longitudinal axis. Figure 6.27 shows the value of $|\kappa|L$ that gives minimal spatial hole burning and the corresponding value of P_{min}/P_{max} as a function of κ_g/κ_n. Little spatial hole burning is obtained in all cases and one can see that uniform power results for pure gain coupling. $|\kappa|L$ equals the value $\pi/2$ in this case and both $\Delta\beta_i$ and $\Delta\beta_r$ are zero. It can easily be verified that the solution of the coupled-wave equations then reduces to

$$R^+(z) = \sin\left(\frac{\pi z}{2L}\right); R^-(z) = \cos\left(\frac{\pi z}{2L}\right) \quad (6.14)$$

The longitudinal variation of forward and backward propagating power, as well as the total power, is depicted in Figure 6.28 for 300-μm-long lasers for three values of κ_g/κ_n (10^{-2}, 0.5, and 10). Relatively uniform power is also found for the case of weak gain coupling or almost pure index coupling. The low threshold gain difference in this case makes such a laser unattractive, however. As κ_g increases, both R^+ and R^- become more sinusoidal and the power becomes more uniform.

Figure 6.29 shows the threshold gain difference ΔgL and the output power P_m (the dashed curve on Figure 6.29), where the SMSR drops below 20 dB as a function

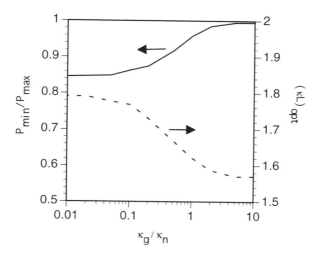

Figure 6.27 Optimum value of P_{min}/P_{max} (solid line) and of κL (dashed line) as a function of κ_g/κ_n.

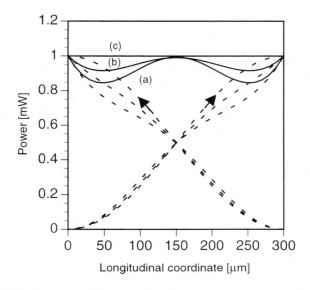

Figure 6.28 Longitudinal variation of the power for κ_g/κ_n.

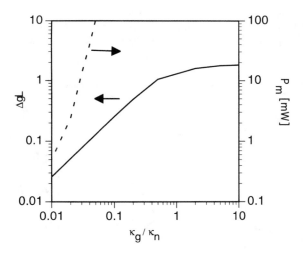

Figure 6.29 Variation of ΔgL (solid line) and P_m (dashed line) vs. κ_g/κ_n.

of κ_g/κ_n. Both ΔgL and P_m increase rapidly for increasing gain coupling. For example, for $\kappa_g/\kappa_n = 0.05$, genuine single-mode behavior (SMSR > 30 dB) is obtained up to an output power level of more than 50 mW.

6.5 THE WAVELENGTH TUNABILITY OF DFB LASERS

The emission wavelength of single-section DFB lasers usually changes continuously as the output power is varied. This wavelength variation has its origin in changes in the average refractive index caused by heating or by changes in the average carrier density due to spatial hole burning or gain suppression. A simplified explanation is as follows. If the laser structure is fixed, resonance is found for a fixed average mode propagation constant β ($\sim n_e/\lambda$) so that any change in the refractive index implies a proportional shift in the emission wavelength.

The power variations that accompany the tuning in single-section lasers are, however, unacceptable for most applications. Moreover, the thermally induced tuning is rather slow while the electronic tuning range is rather small due to the clamping of the gain in such devices. In contrast, a large tuning range under nearly constant output power can be achieved with different types of multisection lasers [32]. A discussion of all possible tunable lasers is not within the scope of this book, and here we only intend to briefly introduce the reader to multisection DFB lasers. One generally distinguishes between discrete and continuous tuning. The term *continuous tuning* is used when the wavelength can be swept continuously over a certain range. In reality this often requires the laser to operate in the same cavity mode during the sweep. *Discrete tuning*, on the other hand, refers to the possibility of having laser emission at several discretely spaced wavelengths or wavelength bands. The transitions between different wavelengths or wavelength bands are caused by mode jumps in this case. The term *quasi-continuous tuning* is sometimes used to describe a continuous wavelength range covered by a number of cavity modes. Continuous tuning is required in, for example, coherent communications; discrete tuning is usually sufficient in WDM communications. Tuning remains an area under intense study.

The tuning of a multielectrode DFB laser is based on a modification of the lasing condition by changing the nonuniformity of the injection. A typical tuning characteristic of a multielectrode DFB laser is shown in Figure 6.30 [33]. The typical tuning range is 2 to 3 nm. However, this tuning range has been shown to improve significantly when single-quantum-well or other quantum-well material with a highly sublinear dependence of the gain on the carrier density is used. Owing to this sublinear nature, a large change in the carrier density in one section can be obtained with a small change in injection in the other section. Using this gain lever effect, continuous tuning over 7 nm has been demonstrated [34].

Chirped gratings or gratings with stepwise period variation are beneficial for the tuning range as well. In this case, wavelengths can be selected by selectively pumping

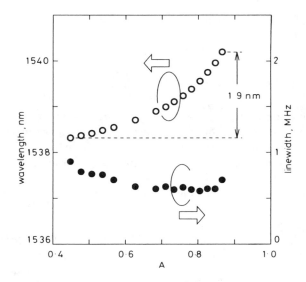

Figure 6.30 Typical tuning characteristic of a multielectrode DFB laser. (*From:* [33].)

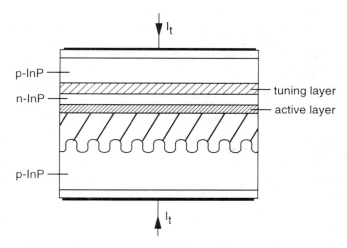

Figure 6.31 Tunable twin-guide DFB laser. (*After:* [37].)

parts where the Bragg wavelength matches the desired emission wavelength. This effect was originally proposed in [35] and has been illustrated both theoretically and experimentally in [36].

An alternative tunable DFB laser is the tunable twin-guide (TTG) DFB laser [37]. The transverse structure of this laser is depicted in Figure 6.31. The active and tuning layers are separated but belong to a single section, and this allows the tuning of the optical length of the DFB laser and the Bragg wavelength of the grating synchronously. An important advantage of the TTG laser is thus the simple tuning scheme based on a single tuning current. The usually shorter cavity length, in addition, allows more lasers to be grown on a wafer. Typical tuning ranges of TTG lasers, which are comparable with those of DBR lasers, are over 4 nm.

6.6 MEASUREMENT OF THE ASE SPECTRUM IN DFB LASERS

A simple and fast method for measuring the spectrum is depicted in Figure 6.32. The laser light is simply launched into an optical fiber that is connected to a commercially available optical spectrum analyzer. The resolution and the dynamic range (i.e., the maximum amplitude ratio between consecutive wavelengths) of the measurements are determined by the type of spectrum analyzer. There are two basic categories: spectrum analyzers based on diffraction grating and on the interferometer. Diffraction grating–based spectrum analyzers, which are the most common today, make use of a finely corrugated mirror to separate the different wavelengths of the incoming light.

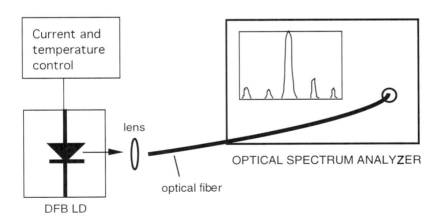

Figure 6.32 Schematic setup for the optical spectrum measurement.

The different wavelengths are consecutively directed toward the photodetector by a computer-controlled rotation of the grating (Figure 6.33).

The wavelength resolution of the instrument is determined by the width of the aperture. Figure 6.32 shows a single monochromator instrument. Instruments with a double monochromator (i.e., with two single monochromators placed in series prior to the photodetector) have a higher dynamic range but a lower sensitivity (the minimum detectable power) due to the insertion loss of the additional optics. A high dynamic range and high sensitivity at the same time only exist for double pass monochromators, in which the light is reflected after the aperture and sent through the monochromator again.

Interferometer-based spectrum analyzers usually have much larger dimensions and therefore have a narrow resolution that is typically specified in a 100-MHz to 10-GHz range. They can also be used to measure the chirp of modulated lasers and, for some Fabry-Perot lasers, even the linewidth. A Fabry-Perot interferometer consists of two highly reflective parallel mirrors that act as a resonant cavity that filters the incoming light (Figure 6.34). The major disadvantage is that multiple wavelengths (with a spacing called the *free spectral range*) are passed by the filter. The

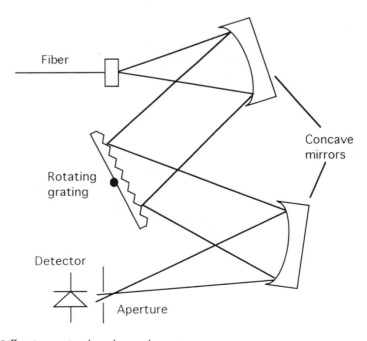

Figure 6.33 Diffraction grating–based monochromator.

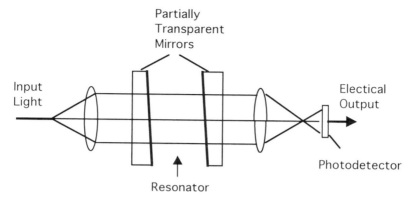

Figure 6.34 Fabry-Perot interferometric optical spectrum analyzer.

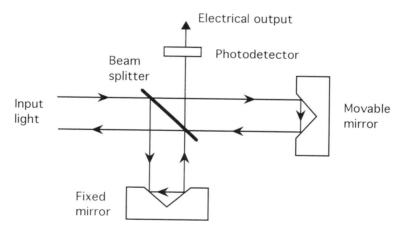

Figure 6.35 Michelson interferometer-based optical spectrum analyzer.

Michelson interferometer (Figure 6.35) creates an interference pattern between the signal and a delayed part of the signal. When the power of the interference pattern is measured as a function of the delay time, a replica of the autocorrelation function of the input signal is obtained and a Fourier transform of this function results in the power spectrum.

6.7 EXTRACTION OF DEVICE PARAMETERS FROM THE SPECTRUM

The expressions derived in (6.1) for the output power as a function of the wavelength are very accurate, analytical, explicit expressions as long as the injected current remains sufficiently below threshold. It should therefore be possible to obtain good agreement between experimentally observed and theoretically calculated spectra, provided the laser parameters are well known. Of these parameters (which include laser length, grating period, and coupling coefficients; complex reflection coefficients of the facets; and effective index and modal gain), only a few (typically the laser length, the grating period, and the amplitudes of the reflection coefficients) are usually known for fabricated lasers.

This lack of parameter values somehow prevents a comparison between theory and experiment, but the expected accuracy of the analytical expressions can, on the other hand, be used to extract some of the unknown laser parameters from the comparison of experimental and theoretical spectra. The method basically consists of minimizing the sum over a number of wavelengths of the squared differences between calculated and experimentally obtained spectral densities using the laser parameters as variables to be optimized. A numerical approach for the optimization using programmed expressions for the spectra is of course needed due to the complexity. This approach, however, also allows one to take into account extra details such as dispersion in refractive index and gain or to apply the method to multisection lasers. Furthermore, it is important to use spectra on a logarithmic scale so that sufficient importance is attached to the stopband, which depends strongly on the coupling constants.

The technique has been demonstrated in [38,39]. An example of the achievable agreement between a measured and a fitted theoretical spectrum is shown in Figure 6.36 [39]. The laser considered here is a third-order gain-coupled DFB laser with cleaved facets, a cavity length of 200 µm, and a grating period of 373.9 nm. The laser parameters obtained from the fitting are summarized in Table 6.2. The span was carefully chosen to include a few Fabry-Perot resonances from which the net gain and group index could be very easily estimated. For these Fabry-Perot resonances, it follows from (6.4) that the ratio r of maximum and minimum is given by

$$r = \frac{1 + \sqrt{R_1 R_2}\, \exp[(\Gamma g - \alpha_{int})L]}{1 - \sqrt{R_1 R_2}\, \exp[(\Gamma g - \alpha_{int})L]} \tag{6.15}$$

while the separation between subsequent maxima (or minima) $\Delta\lambda$ is

$$\Delta\lambda = \frac{\lambda^2}{2 n_g L} \tag{6.16}$$

The Spectrum of DFB Laser Diodes

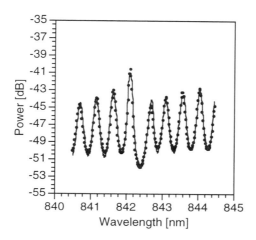

Figure 6.36 Measured (-) and fitted (•) spectra for a cleaved DFB laser.

Table 6.2
Parameter Values Extracted From the Fitting of the Spectrum Shown in Figure 6.36

Parameter	(Unit)	Extracted Value	Definition
κ_n	(cm^{-1})	26.6	Index-coupling coefficient
κ_g	(cm^{-1})	7.3	Gain-coupling coefficient
$\Gamma g_p - \alpha_{int}$	(cm^{-1})	140	Peak gain − absorption
E_p	(eV)	1.467	Peak energy
g_1	(cm^{-1}(eV)$^{-2}$)	−290	Gain curvature (cm^{-1} (eV)$^{-2}$)
n_g		3.72	Group index
$dn_e/d\lambda$	(μm^{-1})	−0.4	Injection efficiency
ϕ_l		−0.7	Grating phase at left-hand-side facet
ϕ_r		−2.8	Grating phase at right-hand-side facet

In the example, a linear dependence on the photon energy was assumed for the refractive index and a quadratic dependence on the photon energy was assumed for the gain. A more accurate wavelength dependence of gain and refractive index, especially in the wavelength region where distributed feedback is important, could be derived from spectra measured on Fabry-Perot lasers made from the same wafer as the DFB laser.

The fitting technique requires rather complicated numerical processing of the experimental spectra, but there is for most lasers no alternative that allows the determination of the coupling coefficients. Index-coupled DFB lasers with both facets highly AR-coated and with uniform grating are an exception. The coupling coefficient of such lasers can be extracted from the width of the stopband.

References

[1] Mols. P.. P. Kuindersma. M. Van Es-Spiekman. and I. Baele. "Yield and Device Characteristics of DFB Lasers: Statistics and Novel Coating Design in Theory and Experiment." *IEEE J. Quant. Electron.*. Vol. 25. June 1989. pp. 1303–1319.
[2] Buus. J.. "Mode Selectivity in DFB Lasers With Cleaved Facets." *Electron. Lett.*. Vol. 21. 1985. pp. 179–180.
[3] Kinoshita. J.. and K. Matsumoto. "Yield Analysis of SLM DFB Lasers With an Axially-Flattened Internal Field." *IEEE J. Quant. Electron.*. Vol. 25. June 1989. pp. 1324–1332.
[4] David. K.. G. Morthier. P. Vankwikelberge. R. Baets. T. Wolf. and B. Borchert. "Gain Coupled DFB Lasers Versus Index Coupled and Phase-Shifted DFB Lasers: A Comparison Based on Spatial Hole Burning Corrected Yield." *IEEE J. Quant. Electron.*. Vol. 27. June 1991. pp. 1714–1723.
[5] Soda. H.. Y. Kotaki. H. Sudo. H. Ishikawa. S. Yamakoshi. and H. Imai. "Stability in Single Longitudinal Mode Operation in GaInAsP/InP Phase-Adjusted DFB Lasers." *IEEE J. Quant. Electron.*. Vol. 23. June 1987. pp. 804–814.
[6] Erman. M.. "Integrated Guided-Wave Optics on III-V Semiconductors." *Proc. 14th International Symp. on Gallium Arsenide and Related Compounds*. Heraklion. Crete. Greece. September 1987. pp. 30–40.
[7] Aoki. M.. K. Uomi. T. Tsuchiya. S. Sasaki. and N. Chinone. "Stabilization of the Longitudinal Mode Against Spatial Hole Burning in $\lambda/4$-Shifted DFB Lasers by Quantum Size Effect." *IEEE Phot. Tech. Lett.*. Vol. 2. September 1990. pp. 617–619.
[8] Schatz. R.. "Longitudinal Spatial Instability in Symmetric Semiconductor Lasers Due to Spatial Hole Burning." *IEEE J. Quant. Electron.*. Vol. 28. June 1992. pp. 1443–1449.
[9] Wünsche. H. J.. U. Bandelow. and H. Wenzel. "Calculation of Combined Lateral and Longitudinal Spatial Hole Burning in $\lambda/4$-Shifted DFB Lasers." *IEEE J. Quant. Electron.*. Vol. 29. 1993. pp. 1751–1760.
[10] Yoshikuni. Y.. K. Oe. G. Motosugi. and T. Matsuoka. "Broad Wavelength Tuning Under Single-Mode Oscillation With a Multi-Electrode Distributed Feedback Laser." *Electron. Lett.*. Vol. 22. 1986. pp. 1153–1154.
[11] Usami. M.. and S. Akiba. "Suppression of Longitudinal Spatial Hole-Burning in $\lambda/4$-Shifted DFB Lasers by Nonuniform Current Distribution." *IEEE J. Quant. Electron.*. Vol. 25. June 1989. pp. 1245–1253.
[12] Kotaki. Y.. S. Ogita. M. Matsuda. Y. Kuwahara. and H. Ishikawa. "Tunable. Narrow-Linewidth and High-Power $\lambda/4$-Shifted DFB Laser." *Electron. Lett.*. Vol. 25. July 1989. pp. 990–991.
[13] Agrawal. G.. J. Geusic and P. Anthony. "Distributed Feedback Lasers With Multiple Phase-Shift Regions." *Appl. Phys. Lett.*. Vol. 53. July 1988. pp. 178–179.
[14] Kimura. T.. and A. Sugimura. "Coupled Phase-Shift Distributed Feedback Semiconductor Lasers for Narrow Linewidth Operation." *IEEE J. Quant. Electron.*. Vol. 25. April 1989. pp. 678–683.
[15] Ogita. S.. Y. Kotaki. M. Matsuda. Y. Kuwahara. and H. Ishikawa. "Long-Cavity. Multi-Phase-Shift. Distributed Feedback Laser for Linewidth Narrowing." *Electron. Lett.*. Vol. 25. 1989. pp. 629–630.
[16] Bissessur. H.. D. Lesterlin. A. Bodéré. E. Gaumont-Goarin. M. Lambert. and C. Duchemin. "Phase-Shifted DFB Laser for Narrow Linewidth Operation." *Proc. ECOC/IOOC 1991*. Paris. France. 1991. pp. 1–4.

[17] Tada. K., Y. Nakano. and A. Ushirokawa. "Proposal of a Distributed Feedback Laser With Nonuniform Stripe Width for Complete Single-Mode Oscillation." *Electron. Lett.*, Vol. 20, 1984, pp. 82–84.
[18] Koyama. F., Y. Suematsu. K. Kojima. and K. Furuya. "1.5μm Phase-Adjusted Active Distributed Reflector Laser for Complete Single-Mode Operation." *Electron. Lett.*, .
[19] Soda. H., K. Wakap. H. Sudo. T. Tanahashi. and H. Imai. "GaInAsP/InP Phase-Adjusted Distributed Feedback Lasers With a Step-Like Nonuniform Stripe Width Structure." *Electron. Lett.*, Vol. 20, 1984. pp. 1016–1018.
[20] Okai. M., T. Tsuchiya. K. Uomi. N. Chinone. and T. Harada. "Corrugation-Pitch-Modulated MQW-DFB Laser With Narrow Spectral Linewidth (170 kHz)." *IEEE Phot. Tech. Lett.*, Vol. 2. August 1990. pp. 529–530.
[21] Zhou. P., and G. S. Lee. "Mode Selection and Spatial Hole Burning Suppression of a Chirped Grating Distributed Feedback Laser." *Appl. Phys. Lett.*, Vol. 56. April 1990. pp. 1400–1402.
[22] Hillmer. H., K. Magari. and Y. Suzuki. "Chirped Gratings for DFB Laser Diodes Using Bent Waveguides." *IEEE Phot. Tech. Lett.*, Vol. 5, 1993. pp. 10–12.
[23] Salzman. J., H. Olesen. A. Møller-Larsen. O. Albrektsen. J. Hanberg. J. Hørregaard. and B. Tromborg. "The S-Bent Waveguide Distributed Feedback Laser." *Proc. 14th International Semiconductor Laser Conference*. Maui. Hawaii. September 1994. pp. 57–58.
[24] Schrans. T., and A. Yariv. "Semiconductor Lasers With Uniform Longitudinal Intensity Distribution." *Appl. Phys. Lett.*, Vol. 56. April 1990. pp. 1526–1528.
[25] Morthier. G., K. David. P. Vankwikelberge. and R. Baets. "A New DFB Laser Diode With Reduced Spatial Hole Burning." *IEEE Phot. Tech. Lett.*, Vol. 2. June 1990. pp. 170–172.
[26] Heise. G., R. Max. and U. Wolff. "Phase-Shifted Holographic Gratings For Distributed Feedback Lasers." *Proc. SPIE Conference on Integrated Optical Circuit Engineering III*. 1986. pp. 87–91.
[27] Talneau. A., J. Charil. A. Ougazzaden. and J. C. Bouley. "High Power Operation of Phase-Shifted DFB Lasers With Amplitude Modulated Coupling Coefficient." *Conference Digest of the 13th IEEE International Semiconductor Laser Conference*. Takamatsu. Kagawa. Japan. September 1992. pp. 218–219.
[28] Morthier. G., P. Vankwikelberge. K. David. and R. Baets. "Improved Performance of AR-Coated DFB Lasers by the Introduction of Gain Coupling." *IEEE Phot. Tech. Lett.*, Vol. 2. March 1990. pp. 170–172.
[29] Kogelnik. H., and C. V. Shank. "Coupled-Wave Theory of Distributed Feedback Lasers." *J. Appl. Phys.*, Vol. 43. 1972. pp. 2327–2335.
[30] Kapon. E., A. Hardy. and A. Katzir. "The Effects of Complex Coupling Coefficients on Distributed Feedback Lasers." *IEEE J. Quant. Electron.*, Vol. 18. 1982. pp. 66–71.
[31] Luo. Y., Y. Nakano. K. Tada. T. Inoue. H. Hosomatsu. and H. Iwaoka. "Gain-Coupled DFB Semiconductor Laser Having Corrugated Active Layer." *Proc. 7th International Conference on Integrated Opt. and Opt. Fiber Communication*. Japan. 1989. pp. 40–41.
[32] Koch. T. L., and U. Koren. "InP-Based Photonic Integrated Circuits." *IEE Proc. Pt. J.* Vol. 138. 1991. pp. 139–147.
[33] Kotaki. Y., S. Ogita. M. Matsuda. Y. Kuwahara. and H. Ishikawa. "Tunable Narrow-Linewidth and High-Power λ/4-Shifted DFB Laser." *Electron. Lett.*, Vol. 25, 1989. pp. 990–992.
[34] Kuindersma. P. I., W. Scheepers. J. M. H. Cnoops. P. J. A. Thijs. G. L. A. Van Den Hofstad. T. Van Dongen. and J. J. Binsma. "Tunable Three-Section. Strained MQW. PA-DFB's With Large Single Mode Tuning Range (72 A) and Narrow Linewidth (Around 1MHz)." *Proc. 12th IEEE International Semiconductor Laser Conference*. Davos. Switzerland. 1990. pp. 248–249.
[35] Schrans. T., M. Mittelstein. and A. Yariv. "Tunable Active Chirped-Corrugation Waveguide Filters." *Appl. Phys. Lett.*, Vol. 55. 1989. pp. 212–214.
[36] Chen. N., Y. Nakano. G. Morthier. and R. Baets. "Enhanced Tunability in Chirped Grating DFB Lasers." *LEOS'94 7th Annual Meeting*. Boston. October 1994.
[37] Illek. S., W. Thulke. C. Schanen. H. Lang. and M. C. Amann. "Over 7 nm (875 GHz) Continuous

Wavelength Tuning by Tunable Twin-Guide (TTG) Laser Diode," *Electron. Lett.*, Vol. 26, 1990, pp. 46–47.

[38] Schatz, R., E. Berglind, and L. Gillner, "Parameter Extraction From DFB Lasers by Means of a Simple Expression for the Spontaneous Emission Spectrum, *IEEE Phot. Tech. Lett.*, Vol. 6, October 1994, pp. 1182–1184.

[39] Morthier, G., R. Baets, and Y. Nakano, "Parameter Extraction From Subthreshold Spectra in Cleaved Gain- and Index-Coupled Distributed Feedback Laser Diodes," *Proc. Optical Fiber Communications Conference*, San Diego, February 1995, p. 309.

Chapter 7

The IM and FM Behavior of DFB Laser Diodes

Lasers offer the advantage that they can be directly modulated. Therefore, the dynamic characteristics of lasers are very important. In this chapter we deal with two important dynamic properties of DFB lasers: the IM and the FM response.

First, we will show how the IM and FM responses can be measured. Next, the IM response is discussed in more detail. It is shown that spatial hole burning has a cutoff frequency of a few gigahertz. For the IM response, spatial hole burning is of little relevance. Therefore, the high-speed behavior can be discussed using the lumped rate equations of Chapter 2. Some important parameters, such as the K-factor and the maximum 3-dB modulation bandwidth, are derived.

Next, we will discuss the FM response and the different carrier-induced contributions to it: the transient contribution, the gain suppression contribution, and the spatial hole burning contribution. The latter contribution cannot be neglected in the FM response. It is shown that the spatial hole burning can either cause a blue or a red shift in the lasing wavelength.

Finally, this chapter also discusses lateral spatial hole burning and carrier transport in quantum-well lasers. Both effects tend to limit the 3-dB modulation bandwidth of lasers.

7.1 MEASURING THE IM RESPONSE OF LASER DIODES

The measurement of the IM of a laser diode is straightforward and illustrated in Figure 7.1. The laser under test receives a bias current on which a modulation signal is imposed through a bias tee. A calibrated lightwave receiver is then used to detect the modulated light and to send the captured modulation signal to a microwave spectrum analyzer—a calibrated electrical receiver. The incoherent lightwave receiver consists of a photodiode and a wideband preamplifier. Through calibration the nonflatness of the frequency response of the test setup can be accounted for in determining the frequency response of the laser diode. The measurement allows us to determine the modulation of the optical power as a function of the modulation drive current for various modulation frequencies.

A magnitude offset will occur if there is any loss due to cables, connectors, or optical isolators. The latter will be needed if the reflections from the optical receiver are not sufficiently low. The sensitivity of lasers to back-reflected light increases as modulation frequency increases. High-speed lasers are particularly susceptible to nonlinear, unrepeatable changes in modulation and noise with reflected light [1]. The result is that optical reflections become critical in high-speed measurements. In practice, a typical Fabry-Perot laser requires more than 30 dB of isolation, while a typical DFB laser needs more than 60 dB of isolation.

By adding an electrical directional coupler, the reflected and incident electrical signals associated with the modulation drive current can be separated (see also Figure 7.1). Measuring the magnitude and phase of the reflected electrical signal allows us to calculate the microwave reflection coefficient. The complex impedance of the laser diode can then be determined if the impedance of the port of the test equipment is known. Note that the main path loss of the coupler also has to be accounted for via calibrations.

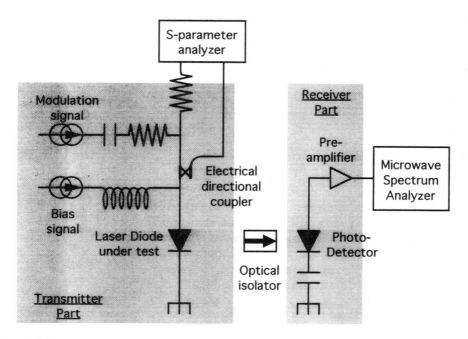

Figure 7.1 IM measurement test setup.

7.2 MEASURING THE FM RESPONSE OF LASER DIODES

Measuring the FM response is somewhat more difficult than measuring the IM response. Several FM measurement approaches are given here. The first approach is based on interferometers, the second on self-homodyning, and the third on fiber dispersion.

7.2.1 FM Measurements Based on Fabry-Perot Interferometers

The approach using scanning *Fabry-Perot interferometers* (FPI) is based on a straightforward measurement of the optical power spectrum [2,3]. In this case, the FPI acts as a filter that only transmits light for which the wavelength matches the cavity length. The transmitted light is detected and shown on an oscilloscope that is synchronized to the scanning frequency of the FPI. In Chapter 6 this setup is discussed with respect to the measurement of the ASE spectrum of DFB lasers.

The resolution of the interferometer is given by $2\Delta f_{1/2}$ = FSR/F, where F and FSR are the finesse and the free spectral range of the FPI, respectively.

If the FM wave is represented by

$$E = E_0 \exp[j\{\omega_0 t + M \sin(\Omega t)\}] \qquad (7.1)$$

then the optical field spectrum of the FM wave can be expanded in terms of Bessel functions $J_i(M)$ of the first kind

$$E = J_0(M) E_0 \sin(j\omega_0 t)$$

$$+ J_1(M) E_0 \sin(j\omega_0 t + j\Omega t) - J_1(M) E_0 \sin(j\omega_0 t - j\Omega t) + \ldots$$

$$+ J_i(M) E_0 \sin(j\omega_0 t + ji\Omega t) + (-1)^i J_i(M) E_0 \sin(j\omega_0 t - ji\Omega t) + \ldots \qquad (7.2)$$

Here, ω_0 is the angular lasing frequency at bias, Ω is the angular modulation frequency, and $M = \Delta\omega_1/\Omega$ is the FM index with $\Delta\omega_1$ the maximum small-signal frequency deviation of the instantaneous angular laser frequency ω. Each side band observed in the measured spectrum corresponds to the square of the coefficient of the corresponding terms in (7.2).

However, this is only true if the modulation frequency is high with respect to the resolution of the FPI. Therefore, the condition $\Omega \gg 2\pi\Delta f_{1/2}$ must apply. In this case, the transmission peak of the FPI is small with respect to the separation of the side bands. Therefore, while scanning across the laser spectrum, only one side band at a time will fall within the peak. Each spectral component is therefore displayed with the correct amplitude relative to the others. The measurement of the FM response

$\Delta\omega_1/I_1 = M\Omega/I_1$ can now be simplified by adjusting the amplitude of the modulation current I_1 in such a way that a known FM index is obtained for the power spectrum of the FM lightwave [4]. Particularly useful FM indexes in this respect are $M = 2.4$ (corresponding to zero amplitude for the carrier frequency in the measured optical spectrum), $M = 1.4$ (corresponding to equal intensity for carrier and first side bands), and $M = 1$ (corresponding to the first side-band intensities in the spectrum that are equal to one-third of the carrier). From the definition of the FM index then follows the FM response $\Delta\omega_1/I_1 = M\Omega/I_1$.

The above approach also requires careful attention to any residual IM. Equation (7.2) is only valid if no spurious IM occurs. However, direct optical FM is unavoidably accompanied by IM and vice versa. Moreover, the influences of IM and FM on the modulated laser spectrum are difficult to separate. The IM index m is defined as $m = \Delta P/P_0$, where P_0 and ΔP, respectively, represent the optical output power at the bias current I_0 and the maximum deviation of the modulated output power. If IM occurs, the power spectrum may be asymmetric with respect to the symmetric FM spectrum expressed in (7.2). However, for $m \ll 1$, the IM can be neglected in the spectrum [3] and the above approach can be applied.

A detailed analysis for the accurate interpretation of FPI measurements is given in [3]. There, attention is also paid to the situation where the assumption $\Omega \gg 2\pi\Delta f_{1/2}$ no longer holds. For the particular case of $\Omega \ll 2\pi\Delta f_{1/2}$, [3] shows that the displayed spectrum no longer reflects the modulated laser spectrum, but a broad spectral envelope that is $2\Delta\omega_1$ wide.

As an alternative to the FPI, a Michelson interferometer (MI) can be used. This method employs the MI as a frequency-to-intensity converter. Analysis of the electrical power spectrum from a detector in the MI's output again yields the FM response [5].

Note that these methods require great care to obtain a stable test environment with a precise optical alignment and a very low level of reflections. These methods are therefore not suitable for performing fast measurements.

7.2.2 The Gated, Delayed Self-Homodyne Technique

The second approach uses a setup as illustrated in Figure 7.2. The laser light is fed into a Mach-Zehnder interferometer that consists of a fiber-optic delay line with delay time T_d and a branch with negligible delay. The modulated DFB laser light is fed into the long delay line. When the modulation has been on during a time T_d, the delay line is full and the modulation is turned off. Unmodulated light coming directly from the DFB laser is now combined with the delayed, modulated light and routed to a photodetector that is connected to an amplifier and an electrical spectrum analyzer. After the modulation has been off for a time T_d, the delay line is full of unmodulated light. Modulation is turned back on, so that modulated light, now coming directly from the laser, is combined with the delayed, unmodulated light.

The photodetector detects the intensity spectrum of the combined waveform,

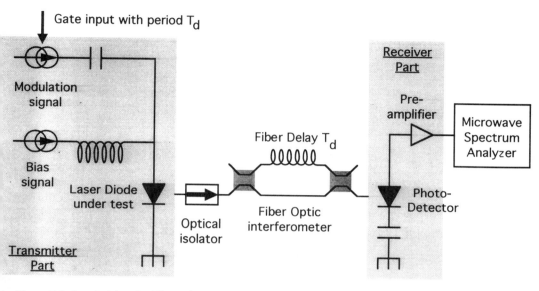

Figure 7.2 Gated, delayed self-homodyne FM measurement test setup.

showing the difference frequency between the modulated light and the unmodulated light. The electrical spectrum analyzer behind the photodetector and amplifier will then display the spectrum of the FM modulation. As described in the previous section, the FM spectrum consists of spectral lines separated in frequency by the modulation frequency $\Omega/2\pi$. The magnitude of the individual spectral lines is again predicted by the appropriate Bessel functions. Note that the width of each spectral component corresponds to the continuous wave or static linewidth of the laser. Therefore, if $\Omega/2\pi$ approaches the linewidth of the laser, the individual components can no longer be resolved.

This measurement method is referred to as self-homodyning because it is based on homodyne detection, with the DFB laser under test also acting as stable local oscillator.

Also, in this method, any spurious IM will be present in the frequency content of the photodiode output. However, since IMs appear as very sharp spectral lines and because electrical spectrum analyzers can have resolution bandwidths below 1 MHz, IM components will show up as very sharp lines while FM components will appear as relatively broad lines (i.e., with the linewidth of the laser). In this way, the FM response can be made visible.

As in Section 7.2.1, the same amplitude adjustments of the modulation current I_1 can be introduced to obtain a power spectrum of the FM light with a known FM index. Unlike the FPI method, no precise optical alignment and no mechanical actuators for moving optical components are required.

7.2.3 Characterization of Laser Chirp Using Fiber Dispersion

The measurement setup of this approach is illustrated in Figure 7.3. A modulated laser diode is used as a light source at the input of the fiber. The IM of the light at the output of the fiber ($IM_{f,out}$) is related to the IM of the laser diode and to the FM of the laser diode. The latter dependence is due to an FM-IM conversion caused by the dispersion of the fiber. Using a small-signal analysis, the corresponding normalized small-signal transfer function of the fiber is given by [6]

$$IM_{f,out} = \cos(F\Omega^2) \, IM_{LD} - (2jP_0/\Omega) \sin(F\Omega^2) \, FM_{LD} \qquad (7.3)$$

where $F = (\lambda^2 D_f L_f)/(4\pi c)$ and λ is the wavelength of the CW laser at bias, L_f is the fiber length, c is the speed of light, D_f is the dispersion coefficient of the fiber, and P_0 is the output power of the CW laser at bias.

The relationship between the FM and the IM response of laser diodes [7],

$$FM_{LD} = \Delta\omega_1/I_1 = \alpha(j\Omega + \Omega_g) \, IM_{LD}/(2P_0) \qquad (7.4)$$

allows us now to express $IM_{f,out}$ in terms of IM_{LD}. The corresponding normalized transfer function looks like

$$H(\Omega) = \cos(F\Omega^2) - (\alpha - j\alpha\Omega_g/\Omega) \sin(F\Omega^2) \qquad (7.5)$$

Equation (7.4) is further explained in Section 7.4.1. Here α is the linewidth enhancement factor and Ω_g is the angular modulation frequency for which the adiabatic and transient chirp of the laser have the same magnitude. In Section 7.4.2, we will see that (7.4) is only applicable to Fabry-Perot lasers. For DFB lasers, (7.4) needs to be extended as shown in (7.23). The corresponding expression for $H(\Omega)$ will also require some changes, but we leave it to the reader to explore this case further.

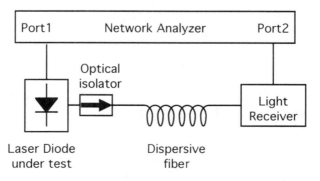

Figure 7.3 Characterization of laser chirp using fiber dispersion.

Using the setup of Figure 7.3, the transfer function of (7.5) can be measured. Curve fitting of the theoretical curve with the experimental curve allows the determination of the three parameters α, D_f (or F), and Ω_g [8]. From the obtained α-parameter and Ω_g, the laser diode FM response can then be determined using (7.4). This is a fast and simple method to measure the chirp of laser diodes [9].

7.3 THE IM RESPONSE

Chapter 2 discusses the small-signal rate equations, and (2.39), in particular, describes the small-signal photon density S_1 and frequency deviation $\Delta\omega_1$ as a function of the angular modulation frequency Ω, the modulation current I_1, the bias value of the photon density, and some laser parameters (gain and losses). These equations represent the intrinsic dynamics of the laser when spatial hole burning is neglected. In the remainder of this chapter we will use as much as possible the simple lumped rate equation model of Chapter 2. This simple model is particularly valid above the cutoff frequency of spatial hole burning. The impact of spatial hole burning on the IM and FM responses will be discussed in the next sections using simulation results obtained with the longitudinal narrowband modeling approach of Chapter 4 (CLADISS).

The IM response is described in (2.39) by the expression for S_1. This expression can be applied to Fabry-Perot and DFB lasers. However, its use for DFB lasers requires some discussion.

7.3.1 The Subgigahertz IM Response

Figure 7.4 illustrates the longitudinal carrier distribution inside a Fabry-Perot laser as a function of the modulation frequency. Although the carrier concentration has a strong longitudinal variation, Figure 7.4 shows that this variation reaches cutoff before the resonance peak occurs. This result was obtained with the longitudinal model of Chapter 4 for a laser without gain saturation. For low modulation frequencies, the model of Chapter 2 would then show the average carrier modulation approaching zero as Ω drops.

However, for FP lasers without gain saturation, the clamping of the average carrier density at low frequencies also follows from the longitudinal model. When the spontaneous emission and noise in the longitudinal static coupled-wave equations are neglected, integration of (3.24) shows that for Fabry-Perot lasers in the static regime the following expression holds

$$\int_0^L 2\,\mathrm{Re}(\Delta\beta)\mathrm{d}z = \frac{1}{2} \ln \frac{1}{R_1 R_2} \tag{7.6}$$

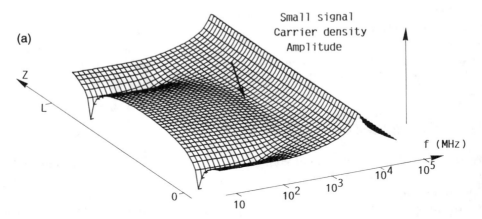

Figure 7.4 (a) Amplitude and (b) phase of the small-signal carrier distribution in the longitudinal direction versus modulation frequency for a Fabry-Perot laser at a bias $I_{bias}/I_{th} = 1.1$.

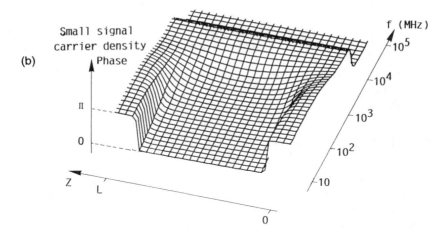

Figure 7.4 (continued)

in which $\Delta\beta$ is approximately linear in $N(z, t)$. Therefore, at low modulation frequencies, the areas in and out of phase with the injected current with, respectively, phase zero and phase π for low Ω in Figure 7.4, must compensate for each other. This also implies that the mirror losses $v_g[s^+(L) - s^-(L) + s^-(0) - s^+(0)]$ of the Fabry-Perot laser are independent of the bias current. Therefore, spatial hole burning will not contribute to the IM response of the Fabry-Perot laser.

For DFB lasers the situation is somewhat more complicated. The main difference between both types of devices is the fact that in DFB lasers the frequency-selective feedback losses depend on the longitudinal distribution of the carriers, which is influenced by spatial hole burning. However, in practice, spatial hole burning has little influence on the IM response of DFB lasers [10] unless the laser becomes unstable and multimodal as illustrated in Chapter 6. In that case, the laser becomes useless anyway. Equations (7.8) and (7.16), discussed further on, justify this assumption.

Figure 7.5 shows the IM response of a simple DFB laser. The subgigahertz response is flat and does not depend on the bias level.

7.3.2 The Spatial Hole Burning Cutoff Frequency

To estimate the cutoff frequency Ω_{SHB} of spatial hole burning, we write the carrier density and the number of photons per unit length as

$$N(z, t) = N_{\mathrm{av}}(t) + N_{\mathrm{SHB}}(t)F(z)$$

$$s(z, t)\frac{L}{V_{\mathrm{act}}} = S(t)[1 + F(z)]$$

$$F = |f^+(z)|^2 + |f^-(z)|^2 - 1$$

$$\int_0^L F(z)\mathrm{d}z = 0 \qquad (7.7)$$

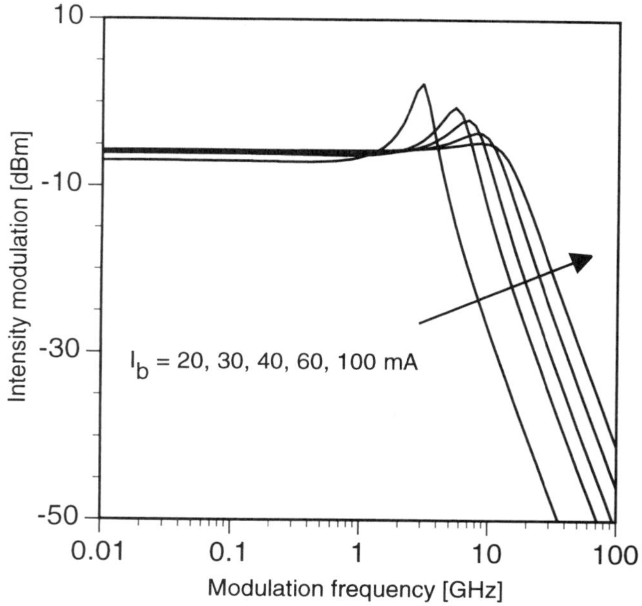

Figure 7.5 Amplitude of the IM response as a function of the modulation frequency for a 300-μm-long DFB laser ($\kappa L = 2$) at different bias levels.

Here S represents the average photon density in the lasing mode. The function F expresses the spatial variation of the carrier density and the photon density along the laser. The functions $f^{\pm}$ are given by (4.26).

Substitution of (7.7) in the carrier rate equation (3.27) and integration over the cavity length L gives the rate equation for $N_{av}(t)$. Since $N_{SHB} \ll N_{av}$, second-order terms in N_{SHB} are neglected. Similarly, substitution of (7.7) in (3.27), followed by multiplication with $F(z)$ and integration over L, gives a rate equation for the spatial hole burning N_{SHB}. We thus get

$$\frac{dN_{av}}{dt} = \frac{\eta J}{qd} - (AN_{av} + BN_{av}^2 + CN_{av}^3) - G(N_{av},S)S$$

$$\frac{dN_{SHB}}{dt} = -\left(A + 2BN_{av} + 3CN_{av}^2 + \frac{\partial G}{\partial N}S\right)N_{SHB} - G(N_{av},S)S \quad (7.8)$$

Introducing the small-signal regime with bias values indicated by the index 0 gives the angular cutoff frequency for spatial hole burning:

$$\Omega_{SHB} \approx A + 2B\,N_{av,0} + 3C\,N_{av,0}^2 + \frac{\partial G}{\partial N}S_0 \quad (7.9)$$

Equation (7.9) indicates that Ω_{SHB} increases for growing bias levels. Calculation shows that the cutoff frequency is on the order of a few gigahertz. Therefore, spatial hole burning has little influence on the resonance peak of the IM response [10]. This resonance peak is characterized by the resonance frequency and the damping rate, which we will now discuss.

7.3.3 The High-Frequency (>GHz) IM Response

In treating the high-speed IM response of Fabry-Perot and DFB lasers, we can limit ourselves to the simpler model of Chapter 2. The nominators in the small-signal equations of (2.39) have the form $\Omega^2 - j\theta\Omega - (2\pi f_r)^2$, corresponding to a second-order resonant system with resonance frequency f_r and damping rate θ as given by (2.40). In the expression for the damping rate, the contribution $R_{sp}/(V_{act}S_0)$ of the spontaneous emission that couples into the mode can be ignored. A few milliwatts above threshold this term is neglectable in single-mode lasers with respect to the more dominant nonlinear gain term $(\partial G/\partial S)S_0$ [11]. According to (2.40), both f_r and θ are then linear functions of the bias photon density S_0.

To determine f_r and θ from experiments, a least-squares fit is generally executed on the IM response. If this is done for different bias points, a plot of θ/S_0 versus $(f_r)^2/S_0$ can be drawn. Measurements show the experimental relationship

$$\theta = K(f_r)^2 \tag{7.10}$$

This linear relationship is also reflected in (2.40), where for sufficiently high power levels the carrier lifetime contribution $(1/\tau_d)$ and the spontaneous emission contribution to θ can be neglected. The value for K then becomes

$$K = \frac{4\pi^2}{G_{th}} \frac{\partial G/\partial N - \partial G/\partial S}{\partial G/\partial N - \partial \gamma/\partial N} \tag{7.11}$$

If K is determined for a single bias point through measurements of f_r and θ, it can be used to determine the damping rate at other bias points from (7.10) and the direct measurement of the resonance frequency itself.

In order to increase the modulation bandwidth of the laser, the resonance frequency can be increased. Equation (2.40) shows that, to this end, the differential gain, the bias power, and the threshold gain can be increased. The latter corresponds to a decreasing photon lifetime. On the other hand, an increasing differential gain and bias power also cause the damping rate to grow. As (2.40) shows, this damping rate will grow even faster with bias photon density if strong gain suppression $\partial G/\partial S$ is present. Therefore, as the power density grows, the resonance frequency goes up, but eventually the damping rate becomes so high that the system becomes overdamped. At that moment the resonance peak disappears and the resonance frequency is no longer a good measure of the modulation bandwidth. The -3-dB bandwidth is then a better measure.

At this point, the reader should take another look at Figure 7.5, which shows the IM response of a DFB laser for different bias levels. For low biases, the laser is underdamped. The intensity equation of (2.39) then shows two complex conjugate poles in the left half of the complex frequency plane. As the bias power increases, the poles move toward the real axis of this plane. Eventually they will reach the real axis and move along it in opposite directions. At the transition point, the system is critically damped as it goes from an underdamped to an overdamped situation. The 3-dB modulation bandwidth will then have reached its maximum.

The angular 3-dB cutoff bandwidth Ω_{3dB} associated with the intrinsic IM response is derived by setting the squared amplitude of the intrinsic IM response of (2.39) to half its low-frequency value. This results in the following equality

$$(\Omega_{3dB}^2 - \Omega_r^2)^2 + \theta^2 \Omega_{3dB}^2 = 2\Omega_r^4 \qquad (7.12)$$

Since the damping rate increases linearly with the bias photon density S_0, the intrinsic resonance will become critically damped at some bias level and the 3-dB bandwidth will then reach its maximum value for that laser diode. This maximum modulation bandwidth is represented by $f_M = \Omega_M/2\pi$. To obtain its value, (7.12) is derived with respect to S_0. Setting $d\Omega_{3dB}/dS_0 = 0$ and using equations (2.40), (7.12), and the approximation (7.10) (which is only valid far enough above threshold as is the case here), then gives

$$f_M = \frac{\Omega_M}{2\pi} = \frac{\sqrt{2}\, 2\pi}{K} \qquad (7.13)$$

This maximum modulation bandwidth is a good indication of the intrinsic maximum speed performance of a laser diode. Typically, Ω_M is on the order of several tens of gigahertz (20 to 40 GHz). Note that, on the other hand, the cutoff frequency for spatial hole burning appears around only a few gigahertz.

7.4 THE FM RESPONSE

In the discussion on the IM response, we treated Fabry-Perot and DFB lasers together, because spatial hole burning has little influence on the IM response. Moreover, the main parameters of interest to both lasers are related to the high-speed characteristics of the lasers where spatial hole burning is no longer relevant. For the FM response, the differences between both types of lasers are too large for this common approach to be maintained.

7.4.1 The FM Response of Fabry-Perot Lasers

Although this book deals with DFB lasers, it is considered justifiable to cover also the FM response of Fabry-Perot lasers in order to gain a better understanding of DFB lasers. The FM response of Fabry-Perot lasers is described in (2.39) by the expression for $\Delta\omega$. Equation (7.4) gives the link between the carrier-induced FM and IM responses. From (2.39), it follows that the adiabatic cutoff frequency Ω_g in (7.4) is given by

$$\Omega_g = -(\partial G/\partial S) S_0 \qquad (7.14)$$

We assumed here that $\partial\gamma/\partial N = 0$ in (2.39). For Ω below Ω_g, the chirp is basically independent of Ω and determined by its low frequency value. This is the adiabatic chirp induced by direct carrier modulation. Above Ω_g the FM response becomes Ω-dependent and the Ω-dependent part is called the *transient chirp*. Equations (7.4) and (7.14) also indicate that the adiabatic FM response increases in magnitude with the bias photon density S_0 and with the level of gain suppression $\partial G/\partial S$.

For very low frequencies, the adiabatic chirp needs to be corrected with a thermally induced chirp that can easily exceed the carrier-induced FM at low frequencies. The thermal FM can be represented by an additional term in (7.4) [7]

$$FM_{LD} = \alpha(j\Omega + \Omega_g) IM_{LD}/(2P_0) + C_{th}/(1 + j\Omega/\Omega_{th}) \qquad (7.15)$$

The thermal effect is represented by a low-pass filter with a thermal cutoff frequency $\Omega_{th}/2\pi$. This frequency is typically on the order of 100 kHz to 1 MHz. For Fabry-Perot lasers, C_{th} is typically negative and several gigahertz/milliamperes in magnitude. Experimentally one also observes that thermal FM is almost bias-independent. This is explained in [12] by observing that the thermal modulation is mainly related to the modulated heat generation in the active region. This generation is almost independent of the bias current due to the clamping of the voltage drop across the DH junction above threshold [13]. From now on, we will neglect this thermally induced FM response, in particular because in DFB lasers the emission frequency of the laser is less sensitive to temperature due to the grating (–20 GHz/K for Fabry-Perot lasers versus –5 GHz/K for DFB lasers).

Figure 7.6 shows the magnitude and phase of the carrier-induced FM response of a Fabry-Perot laser for different bias levels and for different values of the gain suppression $\partial G/\partial S$. The gain suppression causes the low-frequency FM response to be nonzero below the adiabatic cutoff frequency Ω_g. One sees that the FM response drops to zero as Ω goes to zero if no gain suppression occurs ($\Omega_g = 0$). This is in accordance with (7.4), which is based on the lumped rate equations.

Note that although spatial hole burning occurs in the Fabry-Perot laser (see Figure 7.4), it has no contribution to the FM response (see Figure 7.6). This is because spatial hole burning does not affect the gain or the losses and therefore the average carrier density remains clamped to its threshold value when the modulation frequency approaches zero and no gain suppression is considered ($\partial G/\partial S = 0$). On the other hand, if gain suppression occurs, the linear gain (i.e., the S-independent part of the

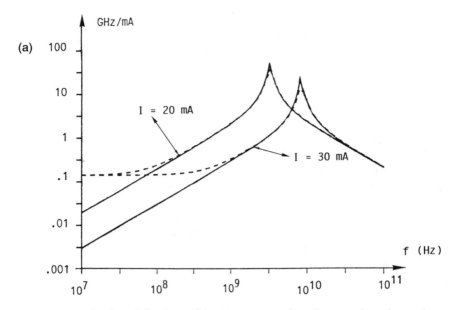

Figure 7.6 (a) Amplitude and (b) phase of the FM response of a Fabry-Perot laser for two bias currents (I_{th} = 18.2 mA). The solid line indicates FM without gain suppression, the dashed line indicates FM with gain suppression.

gain) is not fully clamped at its threshold value, but varies a little with the modulation. Therefore, the carrier concentration, the refractive index, and the wavelength also vary a little with the modulation current.

7.4.2 The FM Response of DFB Lasers

Below we give a simple treatment of the carrier-induced FM response. This description focuses on the different physical phenomena involved. To this end we rewrite the small-signal rate equation for the photon density (see (2.37)) as

$$\left(j\Omega - \frac{\partial G}{\partial S} S_0\right) S_1 = \left(\frac{\partial G}{\partial N} N_{av,1} - \frac{\partial \gamma}{\partial N_{SHB}} N_{SHB,1}\right) S_0 \qquad (7.16)$$

We have assumed here that the carrier density is of the form (7.7) and that N_{SHB} is given by (7.8). The derivatives with respect to N refer here to derivatives with respect to N_{av}. The $\partial \gamma / \partial N_{SHB}$ term in (7.16) arises from a spatial modulation in the carrier density along the cavity. It expresses the dependence of the feedback loss on the level of spatial hole burning.

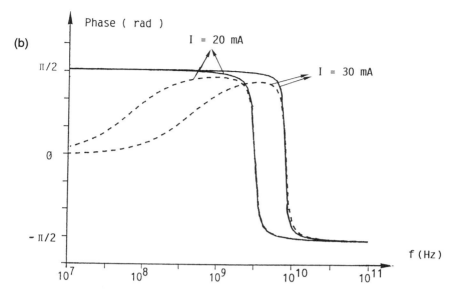

Figure 7.6 (continued)

The small-signal equivalent for the field phase basically expresses that the round-trip phase in the cavity remains constant (modula-2π). It can be cast into an expression that relates the frequency sweep $\Delta\omega_1$ to the real part of an effective refractive index modulation Δn_{re}

$$\frac{\Delta\omega_1}{v_g} = -\frac{\omega}{c}\Delta n_{re} \qquad (7.17)$$

In general, for distributed feedback structures, this effective index modulation Δn_{re} differs from the average effective modal index modulation $\Delta n_{e,av}$ of the waveguide. Hence, Δn_{re} is to be written as the sum of this average effective modal index and a contribution $\Delta n_{e,SHB}$ that represents the direct influence of spatial hole burning on the spectral behavior of the phase of the distributed reflections

$$\Delta n_{re} = \Delta n_{e,av} + \Delta n_{e,SHB} \qquad (7.18)$$

One could compare this way of representation with replacing a change in cavity length by a fictitious change in refractive index.

Based on the above remarks, we can rewrite (7.17) as

$$\frac{\Delta\omega_1}{v_g} = -\frac{\omega}{c}\frac{\partial n_{re}}{\partial N_{av}} N_{av,1} - \frac{\omega}{c}\frac{\partial n_{re}}{\partial N_{SHB}} N_{SHB,1} \qquad (7.19)$$

Index 1 again refers to a small-signal variable, while index 0 corresponds to steady-state bias variables. Substituting now the expression for the linewidth enhancement factor, that is,

$$\alpha = -2\omega \frac{v_g}{c}\left(\frac{\partial n_{re}}{\partial N} \bigg/ \frac{\partial G}{\partial N}\right), \frac{\partial n_{re}}{\partial N} = \frac{\partial n_{e,av}}{\partial N} \qquad (7.20)$$

and a similar expression for the relationship between the spatial hole burning induced loss and index changes in the cavity, i.e.,

$$\psi = 2\omega \frac{v_g}{c}\left(\frac{\partial n_{re}}{\partial N_{SHB}} \bigg/ \frac{\partial \gamma}{\partial N_{SHB}}\right), \frac{\partial n_{re}}{\partial N_{SHB}} = \frac{\partial n_{e,SHB}}{\partial N_{SHB}} \qquad (7.21)$$

into (7.19) yields

$$\Delta\omega_1 = \frac{\alpha}{2}\frac{\partial G}{\partial N} N_{av,1} - \frac{\psi}{2}\frac{\partial \gamma}{\partial N_{SHB}} N_{SHB,1} \qquad (7.22)$$

The second term of the right-hand side of (7.22) only arises when the feedback in the laser is distributed. Unlike α, which is a conceptual "constant," the factor ψ may differ from device to device, depending on details of the distributed feedback. $N_{av,1}$ can now be eliminated from (7.22) by using (7.16). This gives the following expression for $\Delta\omega_1$:

$$\Delta\omega_1 = \frac{\alpha}{2}\left(j\Omega - \frac{\partial G}{\partial S}S_0\right)\frac{S_1}{S_0} + \frac{1}{2}(\alpha - \psi)\frac{\partial \gamma}{\partial N_{SHB}}N_{SHB,1} \qquad (7.23)$$

Equation (7.23) describes the extension of (7.14) when spatial hole burning is included. It shows that there are in principle three physically distinguishable contributions to the FM response.

The first contribution is represented by $\Delta\omega_1 = (\alpha/2) j\Omega S_1/S_0$. It is the ever-present transient contribution. For low frequencies, this term can be neglected, but as Ω nears the resonance frequency, this term becomes the dominant contribution to the carrier-induced FM.

The second contribution, $\Delta\omega_1 = -(\alpha/2)(\partial G/\partial S)S_1/S_0$, is due to gain suppression. This contribution to the FM response is frequency- and bias-independent below the resonance frequency. It should be stressed that spatial hole burning does not contribute to this term (as long as α is carrier concentration–independent). The reason for this is simple if one looks at the physics of gain suppression. Assume for instance that gain suppression is due to spectral hole burning. Spectral hole burning will not give rise to an increasing shortfall of carriers "at the lasing frequency," but instead to an increasing excess of carriers at energies "outside the lasing frequency." Obviously, the number of carriers "at the laser frequency" does not drop, but remains constant, otherwise the device stops lasing. The increasing excess of carriers "outside the laser frequency" leads to a decreasing absorption at other wavelengths. Associated with this lower absorption is a decrease in the real part of the refractive index at the lasing wavelength. The latter can be calculated via, for example, a Kramers-Krönig relation (see Chapter 2).

Similar to spectral hole burning, spatial hole burning does not give rise to an increasing shortfall of carriers "at the laser frequency" when averaged along the laser cavity. For spatial hole burning only, this implies, however, that (on average) there is also no excess of carriers "outside the laser frequency" and hence no associated FM (as in the Fabry-Perot laser), unless spatial hole burning affects the round-trip gain and phase conditions. If it does so, the third contribution arises. With respect to this term it is instructive to consider again a Fabry-Perot laser cavity. Though appreciable spatial hole burning may occur, there is no spatial hole burning contribution to the FM, simply because the feedback loss in such a cavity is fixed.

For DFB lasers the situation is very different. There the feedback is caused by distributed Bragg reflections, which are determined by the precise values of facet reflectivities (modulus and especially phase), by the κL product, and by modulations in the uniformity of the grating (e.g. $\lambda/4$-shifted devices) [14]. Furthermore, the spectral behavior of the distributed feedback is a sensitive function of the internal carrier distribution and therefore also of the power distribution in the laser (see also Chapter 6). Hence, when the power level changes with direct modulation, the amount of spatial hole burning changes. This in turn causes a shift in wavelength because spatial hole burning affects the round-trip phase and the feedback losses, and consequently also the gain and carrier concentration. So, basically, there are two contributions to this wavelength shift: a direct one related via ψ to the round-trip phase condition and an indirect one related via α to the unity round-trip gain condition. It should be understood that these two spatial hole burning contributions to the shift in wavelength are interdependent. The last term in (7.23) illustrates this.

At low frequencies the spatial hole burning FM contribution may be in phase or out of phase with the gain suppression contribution. At high frequencies (above Ω_{SHB}), the spatial hole burning will roll off (in contrast to the fast gain suppression process). In general, the loss in DFB devices is affected by spatial hole burning in a nonlinear way. Therefore, the spatial hole burning contribution is bias-dependent, in contrast with the bias-independent gain suppression FM.

To illustrate the FM response of a DFB laser, calculation results are presented for two different DFB lasers whose device parameters are given in [4]. The two devices have been selected from a large set of calculations with CLADISS on many DFB lasers, and they clearly exemplify FM behavior. Both devices have a κL of 2 and 32% back-facet power reflectivity. One device (labeled A) has an ideal AR coating on the front facet, whereas the other (labeled B) has a 5% front-facet power reflectivity. The choice of the facet phases is random, except that both devices were checked to show stable single-mode behavior up to high bias levels.

For both devices, the FM response is calculated with spatial hole burning always present. In order to emphasize the effect of spatial hole burning, calculations with and without gain suppression are done. Table 7.1 shows the different cases (with and without gain suppression) considered. Gain suppression is expressed here by $(1 - \varepsilon P)$, where P expresses the average power. It is related to (2.32) via $\xi = \varepsilon h v_g V_{act}/(\Gamma L)$.

The small-signal carrier density modulation $N_1(z, \Omega)$ for device B at a bias of 30 mA is depicted in Figure 7.7. Gain suppression has a negligible influence on the spatial distributions of $N_1(z, \Omega)$ [4].

Despite the large similarity in the spatial hole burning component of the carrier density modulation in devices A (not shown) and B (Figure 7.7), the calculations

Table 7.1
Parameters That Differentiate the Examples in Section 7.4.2

Case	R_1	ϕ_1	R_2	ϕ_2	ε (W^{-1})	$\Delta\alpha L$
A_0	0.32	$-\pi/4$	0	—	0	0.32
A_1	0.32	$-\pi/4$	0	—	1	0.32
B_0	0.32	$\pi/2$	0.05	0	0	0.28
B_1	0.32	$\pi/2$	0.05	0	1	0.28

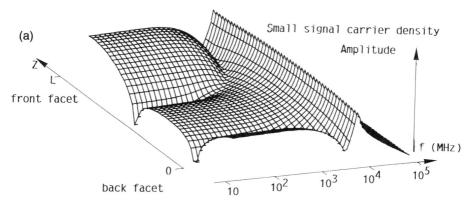

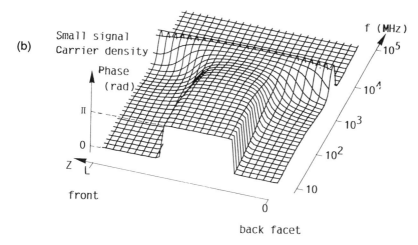

Figure 7.7 (a) Amplitude and (b) phase of the small-signal carrier density distribution in the longitudinal direction versus modulation frequency for DFB laser B_0 ($R_1 = 5\%$, $R_2 = 32\%$, no gain suppression) for a bias $I_{bias} = 1.4 I_{th}$.

Figure 7.7 (continued)

show that the spatial hole burning contribution to the FM response is totally different for the two DFB laser examples. Leaving out the gain suppression for a moment (i.e., $\varepsilon = 0 W^{-1}$), the FM responses for devices A_0 and B_0 as a function of modulation frequency $f(=\Omega/2\pi)$ at various bias levels are shown in Figures 7.8 and 7.9. Devices A_0 and B_0 have in common a flat FM response at low frequencies due to spatial hole burning. The level of this response decreases with increasing bias.

As shown in Figures 7.8 and 7.9, the striking difference between A_0 and B_0 is the phase of the FM at low and intermediate frequencies. At low frequency, the FM phase of A_0 is π (indicating a red shift), whereas for B_0 it is 0 (indicating a blue shift).

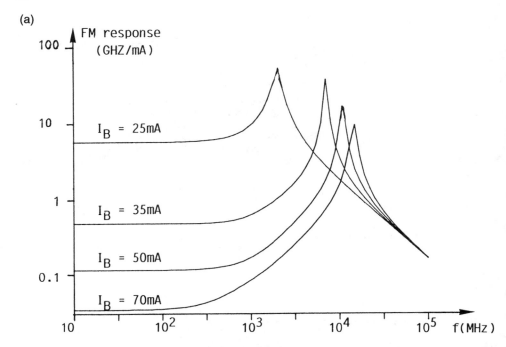

Figure 7.8 (a) Amplitude and (b) phase of the FM response of device A_0 ($R_1 = 0\%$, $R_2 = 32\%$, no gain suppression) for several bias currents.

A pronounced difference in the amplitude of the FM versus frequency is associated with this difference in phase. Device A_0 shows a gradual increase around the gigahertz range, and then a further increase into the relaxation oscillation. However, for device B_0, a pronounced dip appears in the FM versus frequency around the gigahertz range before the relaxation oscillation takes over. Obviously, the dip is caused by the roll-off of the spatial hole burning at high frequency. Simultaneously with this dynamic roll-off around Ω_{SHB}, the phase of the spatial hole burning FM component will quite abruptly change either from π to $\pi/2$ (case A) or from 0 to $-\pi/2$ (case B). It is then the compound effect of the ever-present transient term, which increases with Ω and has phase $\pi/2$ below f_r, and the spatial hole burning component that determines whether or not a dip in the FM amplitude versus frequency will occur. When at roll-off both components have phase $\pi/2$, a gradual increase will occur (case A). However, when the phase of the terms becomes $\pi/2$ (transient component) and $-\pi/2$, respectively, then a dip will appear (case B).

Simultaneous gain suppression and spatial hole burning further complicate the FM response, particularly for DFB device A (now A_1). The consequences for B (now B_1) are less pronounced. We assume here a gain suppression of $\varepsilon = 1W^{-1}$. The calculated FM responses of A_1 and B_1 as a function of frequency at various bias levels are given in Figures 7.10 and 7.11, respectively.

(b)

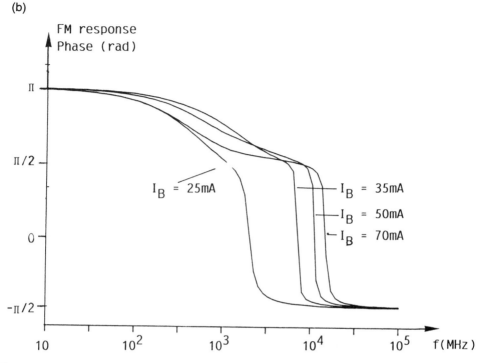

Figure 7.8 (continued)

A common feature of the FM response of A_1 and B_1 is that at low to intermediate modulation frequencies, the FM response, with increasing bias, now levels off to a constant value (i.e., the gain suppression contribution). Upon comparing A_0 and B_0 with A_1 and B_1, respectively, it can also be observed that at low bias levels (a few milliamperes above threshold) spatial hole burning dominates the low- and intermediate-frequency FM response.

Dominating spatial hole burning FM at low bias, combined with dominating gain suppression FM at high bias, produces few more consequences for device B_1. It only weakens the spatial hole burning dip in the FM versus frequency characteristic around the angular cutoff frequency Ω_{SHB}. However, for device A_1, the two contributions apparently cancel out almost exactly at one specific bias level (50 mA for A_1). As seen from Figure 7.10, at this bias level, the low- and intermediate-frequency FM is proportional to the frequency, and the FM phase is close to $\pi/2$.

The large difference in FM as a function of bias between A_1 and B_1 is easy to understand. The phase of the low-frequency gain suppression FM is equal to zero. Hence device B_1 has the phases "gain suppression 0"/"spatial hole burning 0," whereas device A_1 has the phases "gain suppression 0"/"spatial hole burning π." Therefore,

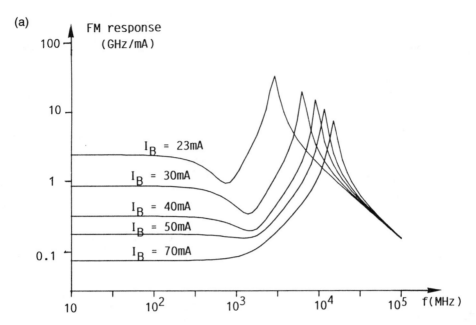

Figure 7.9 (a) Amplitude and (b) phase of the FM response of device B_0 (R_1 = 5%, R_2 = 32%, no gain suppression) for several bias currents.

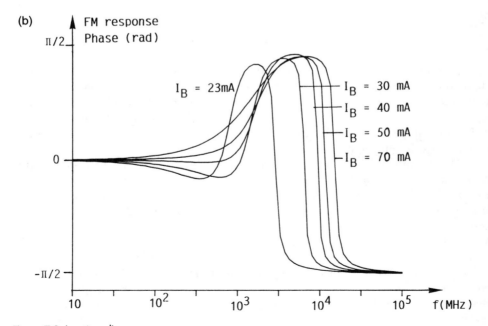

Figure 7.9 (continued)

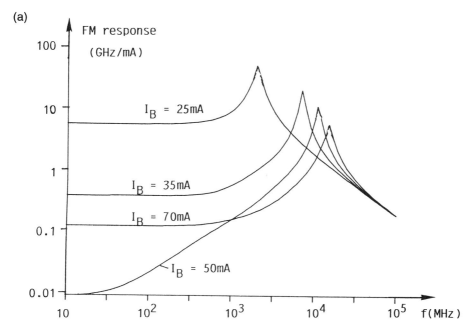

Figure 7.10 (a) Amplitude and (b) phase of the FM response of device A_1 ($R_1 = 0\%$, $R_2 = 32\%$, no gain suppression) for several bias currents.

only in device A_1, can the spatial hole burning FM, which is a decreasing function of the bias level, be compensated for, at one specific bias level, by a constant gain suppression FM. At this bias level, only the transient contribution to the FM is retained, and a pronounced dip will occur in the FM versus bias.

The above calculated effects have been observed experimentally [4,15]. However, the gross features of experimental results are rather characteristic of a dominating gain suppression contribution to FM. Theory does not exclude this possibility, since the calculation used small ε values ($1W^{-1}$) to clearly expose spatial hole burning, whereas typically ε values of 5 W^{-1} to 9W^{-1} are observed.

The thermal FM contribution will also interfere with the spatial hole burning FM and the gain suppression FM. The thermal FM and the gain suppression FM are out of phase. Since the magnitude of the low-frequency thermal contribution is always larger than the gain suppression one, this would lead to a dip in the FM response somewhere at the thermal cutoff. However, this dip will disappear if there is a large enough spatial hole burning contribution, which by chance is also out of phase with the gain suppression one, and hence in phase with the thermal one.

The IM and FM responses of $\lambda/4$ phase-shifted DFB lasers show a behavior

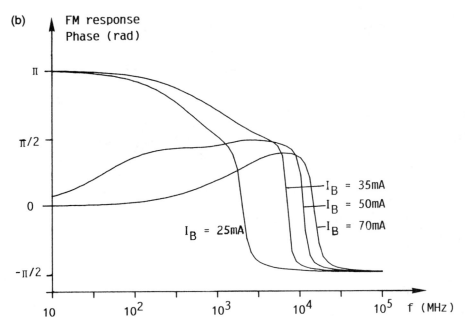

Figure 7.10 (continued)

similar to that of simple DFB lasers. The FM response especially can also be decomposed into the different contributions just discussed, that is, transient FM, spatial hole burning FM, gain suppression FM, and thermal FM [16]. However, calculations indicate that λ/4 phase-shifted DFB lasers with good AR coatings on both facets seem to have a spatial hole burning FM that is very small. Since the facet reflectivities now have a weak impact (ideally none at all), the behavior of such devices is more deterministic.

7.5 LATERAL SPATIAL HOLE BURNING

In Chapter 3 we decided to neglect any lateral variations in the carrier density under the assumption that the width of the active layer (1 to 2 μm) is usually smaller than the diffusion length L_D of the carriers (3 to 5 μm). In devices where this is no longer the case, it becomes relevant to include the lateral diffusion. Some authors [17,18] have studied this situation using lumped rate equations. The derivation of these rate equations is based on expanding the lateral photon density distribution and carrier density distribution in cosine series, from which only the first two terms are retained. This gives

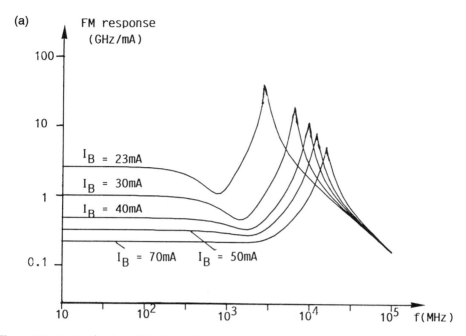

Figure 7.11 (a) Amplitude and (b) phase of the FM response of device B_1 ($R_1 = 5\%$, $R_2 = 32\%$, no gain suppression) for several bias currents.

$$S_y(y, t) = S(t) \frac{1}{w} \left(1 + \cos \frac{2\pi y}{w}\right)$$

$$N_y(y, t) = N(t) - N_l(t) \cos \frac{2\pi y}{w} \quad (7.24)$$

in which $S(t)$ is the average photon density in the laser and $N(t)$ the average carrier density in the active layer. The representation of the lateral carrier distribution in (7.24) expresses the lateral spatial hole burning caused by the lateral variation of the stimulated recombination associated with the lateral field distribution.

Together with the three time variables $S(t)$, $N(t)$, and $N_l(t)$, three rate equations can be derived. With these rate equations, it can be shown that the lateral diffusion contributes to the damping rate. For narrow stripe lasers, this effect may be described by an additional gain compression coefficient as in (2.32) with [17]

$$\xi = \frac{1}{2} \frac{\partial G}{\partial N} \frac{\tau_d}{1 + (2\pi L_D/w)^2} \quad (7.25)$$

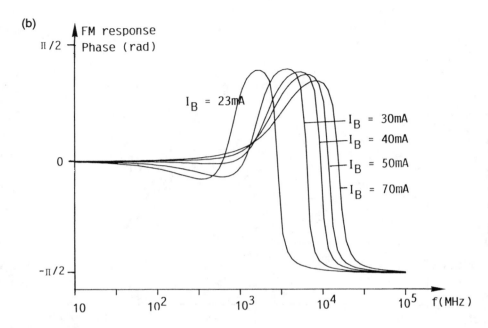

Figure 7.11 (continued)

Assuming $L_D = 3$ μm, $w = 2$ μm, $\partial G/\partial N = 1.46 \times 10^9$ cm³/s, and $\tau_d = 2$ ns, (7.25) yields $\xi = 4.99 \times 10^{-18}$ cm³ (or $\varepsilon \approx 1.1$ W^{-1}) [7]. This value is below the gain compression values due to nonlinear gain phenomena (with ξ values ranging from 1.7×10^{-17} cm³ to 3×10^{-17} cm³ or ε values from $5W^{-1}$ to $9W^{-1}$). In narrow stripe InGaAsP lasers, one can therefore neglect lateral diffusion without too much error.

7.6 DYNAMICS OF QUANTUM-WELL LASERS

Lasers with quantum-well active areas have been shown to have enhanced differential gains with respect to bulk lasers. More improvements in the differential gain can be obtained through the introduction of strain or through modulation doping. The expressions for f_r and f_M then predict a high intrinsic bandwidth compared to bulk lasers.

However, whereas for bulk lasers, K-values are reported in a narrow range of 0.2 to 0.4 ns, reported K-values for quantum-well lasers range from 0.13 to 2.4 ns. Those higher K-values represent lower modulation bandwidths than expected.

A typical *separate confinement heterostructure* (SCH) with a *single quantum well* (SQW) is shown in Figure 7.12. The dynamic behavior of the carrier transport through such an SCH and of the interaction between the SCH and the quantum well can be described concisely by the following carrier density equations [19]

$$\frac{dN_B}{dt} = \frac{\eta I}{qV_{SCH}} - \frac{N_B}{\tau_s} + \frac{N_W}{\tau_e}\frac{V_W}{V_{SCH}}$$

$$\frac{dN_W}{dt} = \frac{N_B}{\tau_s}\frac{V_{SCH}}{V_W} - \frac{N_W}{\tau_d} - \frac{N_W}{\tau_e} - G(N_W, S)S \qquad (7.26)$$

where N_B is the bulk carrier density in the SCH, N_W the carrier density in the quantum-well layer, τ_d the recombination lifetime, V_W the volume of the quantum well, V_{SCH} the volume of the SCH, τ_e the thermionic emission lifetime or escape time (describing the loss of carriers from the quantum well into the SCH), and τ_s the diffusion time (describing the loss of carriers from the SCH into the quantum well, including the ambipolar carrier diffusion transport through the SCH and the local carrier capture time of the well). The photon density rate equation (2.3) is still valid, but in the gain coefficient G, the traditional bulk density N is replaced by the carrier density N_W in the well. Equations similar to (7.26) exist for MQW lasers.

It can now be shown that the inability of the carriers in the SCH to relax to the energy levels of the wells of the SQW or MQW at a rate that is as fast as that of the carrier removal by stimulated recombination influences the laser modulation bandwidth via a capacitive-like roll-off and via a modified K-factor [20].

Indeed, applying a small-signal analysis to (7.26) and (2.3) gives the following approximate expression for the IM response [21]

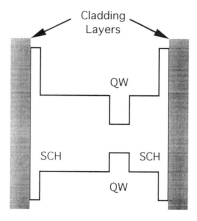

Figure 7.12 Typical band structure of an SCH (multi-) quantum-well laser.

$$\text{IM}_{\text{QWLD}} = \frac{1}{1 + j\Omega\tau_s} \text{IM}_{\text{BLD}}\left(\frac{\partial G}{\partial N} \to \frac{\partial G}{\partial N}\frac{1}{\chi}, \tau_d \to \tau_d\chi\right)$$

$$\chi = 1 + \tau_s/\tau_e \tag{7.27}$$

For $\tau_s = 0$, this equation reduces to the IM response (2.39) for bulk lasers, which is here represented by the function IM_{BLD}.

The first factor in (7.27) is the differential transport factor $\partial I_W/\partial I_S$, where I_W and I_S, respectively, represent the current flowing into the well and the SCH. The dynamics of this factor are determined by the transport time across the SCH. It causes a capacitive-like modulation roll-off that can seriously limit the 3-dB modulation bandwidth of the laser.

Furthermore, not all carriers that flow into the well are effectively kept in it. Some carriers will escape, and this effect is accounted for by the transport factor χ. This means that the K-factor as defined in (7.11), now becomes (neglecting $\partial\gamma/\partial N$)

$$K = (4\pi^2/G_{\text{th}})\cdot(1 - \chi(\partial G/\partial S)/(\partial G/\partial N)) \tag{7.28}$$

Although the quantum well may deliver a higher differential gain $\partial G/\partial N$, this advantage may be offset by the transport factor χ (>1). Obtaining χ-values close to unity requires $\tau_s \ll \tau_e$. Consequently, a narrow SCH is desirable, but this may reduce the optical confinement, leading to a higher threshold carrier density. Due to the rapid gain saturation in quantum-well lasers, this may in turn result in a lower differential gain. Therefore, the SCH width must be optimized for a maximum $(\partial G/\partial N)/\chi$. At the same time, τ_e should also be maximized by optimizing the depth of the wells and the width of the barriers (in case of MQWs).

Due to the rapid gain saturation with increasing carrier density, the threshold gain G cannot be increased freely to reduce K. Here an optimum cavity loss must also be determined by trading off a high differential gain against a high threshold gain. For DFB lasers, this compromise has its impact on the choice of κL. A further complication is the need to optimize κL for low longitudinal spatial hole burning.

Figure 7.13 shows the amplitude modulation of a gain-coupled MQW DFB laser with four 8-nm wells and a 300-nm-wide SCH region and with all other transport parameters taken from [20]. The wide confinement region of this example causes a capacitive-like roll-off at high frequencies and an enhanced damping. The roll-off becomes visible at high power, where the modulation response extends to high frequencies. For wider separate confinement layers, the carrier transport through the confinement structure will take more time (i.e., higher τ_s) and a more pronounced capacitive-like roll-off will occur.

The inability of carriers to relax into the wells is also referred to as the *carrier injection bottleneck* [20]. As a result, carriers tend to accumulate in the energy levels

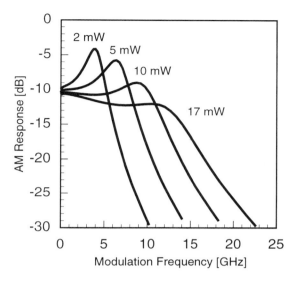

Figure 7.13 Amplitude modulation response of a gain-coupled MQW DFB laser with four 8-nm wells and a 300-nm-wide SCH region. For high bias levels, the capacitive-like roll-off becomes clearly visible.

above the wells. These carriers reside in the barriers and in the SCH (excluding the wells) and strongly influence the real part of the refractive index of those layers. When the laser is modulated, these carriers will also cause a strong refractive index modulation with a strong associated FM contribution. The above discussion lets us assume that a wider SCH will therefore lead to higher FM efficiency. Reference [22] confirms this experimentally and theoretically.

7.7 Designing High-Speed DFB Lasers

Equation (7.13) indicates that to increase the high-speed performance of a laser, the K-parameter of (7.10) should be decreased. To this end, lasers should be designed to operate at high photon densities with high threshold gain (i.e., short photon lifetimes), high differential gain $\partial G/\partial N$, low nonlinear gain $\partial G/\partial S$, and low nonlinear losses $\partial \gamma/\partial N$. In quantum-well lasers especially, extra flexibility is available to adjust these parameters by varying the layer thicknesses, by introducing strain, or by using gain levering [23].

Although a low nonlinear gain favors a low K-factor, high nonlinear gain favors high FM efficiency. Similarly, the carrier transport effect in a wide SCH SQW or MQW laser leads to an extra FM contribution; but on the other hand, this effect can seriously deteriorate the high-speed modulation bandwidth of the laser. Therefore, if a laser is to be used for FM applications, the FM efficiency must be traded off against the high-speed modulation bandwidth. Furthermore, it should be noted that DFB lasers with minimized spatial hole burning are preferred in order to obtain a simpler FM behavior dominated by gain suppression FM.

In practice the maximum modulation bandwidth is often limited by parasitics, device heating, and the maximum power handling capabilities of the laser. This chapter only deals with the dynamics of the physical phenomena in the active layer directly related to the laser action. The active layer is, however, embedded in a semiconductor structure that can show deficiencies. As discussed in Chapter 5, leakage currents can occur around and across the buried heterojunction, and junction and shunt capacitances may affect the dynamics [24]. Also, the bond wire and the package with the laser mount cause additional undesired inductive and capacitive effects.

References

[1] Tkach, R. W., and A. R. Chraplyvy, "Regimes of Feedback Effects in 1.5 μm Distributed Feedback Lasers," *J. Lightwave Tech.*, Vol. LT-4, No. 11, November 1986, pp. 1655–1661.
[2] Kobayashi, S., Y. Yamamoto, M. Ito, and T. Kimura, "Direct Frequency Modulation in AlGaAs Semiconductor Lasers," *IEEE J. Quant. Electron.*, Vol. QE-18, No. 4, April 1982, pp. 582–595.
[3] Olesen, H., and G. Jacobsen, "A Theoretical and Experimental Analysis of Modulated Laser Fields and Power Spectra," *IEEE J. Quant. Electron.*, Vol. QE-18, No. 12, December 1982, pp. 2069–2080.
[4] Vankwikelberge, P., F. Buytaert, A. Franchois, R. Baets, P. I. Kuindersma, and C. W. Fredriksz, "Analysis of the Carrier-Induced FM Response of DFB Lasers: Theoretical and Experimental Case Studies," *IEEE J. Quant. Electron.*, Vol. 25, No. 11, November 1989, pp. 2239–2254.
[5] Eichen, E., and A. Siletti, "The FM Sideband Techniques for Measuring the Frequency Response of Ultra-High Frequency Optical Detectors," *Proc. Optical Fibre Communications Conference*, Reno, paper MJ4, 19–22 January 1987.
[6] Wang, J., and K. Petermann, "Small Signal Analysis for Dispersive Optical Fiber Communication Systems," *IEEE J. Lightwave Tech.*, Vol. 10, No. 1, January 1992, pp. 96–100.
[7] Petermann, K., *Laser Diode Modulation and Noise*, Dordrecht: Kluwer Academic Publishers, 1988.
[8] Srinivasan, R. C., and J. C. Cartledge, "On Using Fiber Transfer Functions to Characterize Laser Chirp and Fiber Dispersion," *IEEE Phot. Tech. Lett.*, Vol. 7, No. 11, November 1995, pp. 1327–1329.
[9] Devaux, F., Y. Sorel, and J. F. Kerdiles, "Simple Measurement of Fiber Dispersion and of Chirp Parameters of Intensity Modulated Light Emitter," *IEEE J. Lightwave Tech.*, Vol. 11, No. 12, December 1993, pp. 1937–1940.
[10] Schatz, R., "Dynamics of Spatial Hole Burning Effects in DFB Lasers," *IEEE J. Quant. Electron.*, Vol. 31, 1995, pp. 1981–1993.
[11] Olshansky, R., P. Hill, V. Lanzisera, and W. Powazinik, "Frequency Response of 1.3 μm InGaAsP High Speed Semiconductor Lasers," *IEEE J. Quant. Electron.*, Vol. QE-23, No. 9, September 1987, pp. 1410–1418.
[12] Kobayashi, S., Y. Yamamoto, M. Ito, and T. Kimura, "Direct Frequency Modulation in AlGaAs Semiconductor Lasers," *IEEE J. Quant. Electron.*, Vol. QE-18, No. 4, April 1982, pp. 582–595.
[13] Joyce, W. B., "Analytic Approximations for the Fermi Energy in (Al,GA)As," *Appl. Phys. Lett.*, Vol. 32, No. 10, 15 May 1978, pp. 680–681.
[14] Haus, A. H., and C. V. Shank, "Antisymmetric Taper of Distributed Feedback Lasers," *IEEE J. Quant. Electron.*, Vol. QE-12, No. 9, September 1976, pp. 532–539.
[15] Kuo, C. Y., M. S. Lin, S. J. Wang, D. A. Ackerman, and L. J. P Ketelsen, "Static and Dynamic Char-

acteristics of DFB Lasers With Longitudinal Nonuniformity," *IEEE Photon. Tech. Lett.*, Vol. 2, No. 7, July 1990, pp. 461–463.

[16] Whiteaway, J. E. A., B. Garrett, G. H. B Thompson, A. J. Collar, C. J. Armistead, and M. J. Fice, "The Static and Dynamic Characteristics of Single Multiple Phase-Shifted DFB Laser Structures," *IEEE J. Quant. Electron.*, Vol. 28, No. 5, May 1992, pp. 1277–1293.

[17] Tucker, R., and D. Pope, "Circuit Modeling of the Effect of Diffusion Damping in a Narrow-Stripe Semiconductor Laser," *IEEE J. Quant. Electron.*, Vol. QE-19, No. 7, July 1983, pp. 1179–1183.

[18] Leclerc, D., P. Brosson, B. Fernier, and J. Benoit, "Relative Contribution of Carrier Diffusion and Spontaneous Emission to the Strength and Damping of Relaxation Oscillations in InGaAsP/InP Lasers," *IEE Proc.*, Vol. 132, Pt. J, No. 1, February 1985, pp. 28–33.

[19] Nagarajan, R., T. Fukushima, S. W. Corzine, and J. E. Bowers, "Effects of Carrier Transport in High Speed Quantum Well Lasers," *Appl. Phys. Lett.*, Vol. 59, 1991, pp. 1835–1837.

[20] Tessler, N., R. Nagar, and G. Eisenstein, "Structure Dependent Modulation Responses in Quantum-Well Lasers," *IEEE J. Quant. Electron.*, Vol. 28, No. 10, October 1992, pp. 2242–2250.

[21] Nagarajan, R., M. Ishikawa, T. Fukushima, R. S. Geels, and J. Bowers, "High Speed Quantum-Well Lasers and Carrier Transport Effects," *IEEE J. Quant. Electron.*, Vol. 28, No. 10, October 1992, pp. 1990–2008.

[22] Yamazaki, H., M. Yamaguchi, M. Kitamura, and I. Mito, "Analysis on FM Efficiency of InGaAs/InGaAsP SCH-MQW LD's Taking Injection Carrier Transport Into Account," *IEEE Photon. Tech. Lett.*, Vol. 4, No. 4, April 1993, pp. 396–398.

[23] Lau, K. Y., "Dynamics of Quantum Well Lasers," Chapter 5 in *Quantum Well Lasers*, P. S. Zory, ed., San Diego: Academic Press, 1993, pp. 217–276.

[24] Lau, K., and A. Yariv, "High-Frequency Current Modulation of Semiconductor Injection Lasers," *Semiconductors and Semimetals, Vol. 22, Lightwave Communications Technology*, W. Tsang, Volume ed., Pt. B, "Semiconductor Injection Lasers I," Chapter 2, Orlando: Academic Press, 1985.

Chapter 8

Harmonic and Intermodulation Distortion in DFB Laser Diodes

It is well known that the performance of analog systems depends on the linearity of the system and that any deviation from the linearity results in distortions of the waveform of the signal and reduces the quality of the received signal. This is no different in optical systems, where it is mainly the laser diode that exhibits a nonlinear behavior. Examples of such optical communication schemes are the SCM systems, which are based on analog intensity modulation of the carriers and are widely applied in CATV distribution and wireless telephony. In CATV applications, there are typically over 50 channels with 5-MHz bandwidth, each in a frequency range from 50 MHz to 1.8 GHz [1,2]. In wireless telephony, SCM is applied on the fiber-optic link between base and central stations [3]. DFB lasers are the desired transmitters for such systems, since they usually also have small intensity noise.

The nonlinearity of the laser diode results in an output power that is not perfectly linear as a function of the injected current, and the nonlinearity is—for sinusoidal small-signal modulation—expressed by the harmonic and intermodulation distortion. The harmonic distortion describes the output at harmonics of the modulation frequency (with the nth-order harmonic distortion describing the output at the nth harmonic), while the intermodulation distortion describes the output at the sum and difference frequencies when the current modulation contains several frequencies. Both the harmonic and the intermodulation distortion lead to an influence of the signal of one carrier or channel on the signal of other carriers or channels. Second- and third-order harmonic distortions are often related to the carrier level, and the ratios are usually abbreviated as 2HD/C and 3HD/C.

The causes of the nonlinear behavior of a DFB laser have been discussed in the previous chapters and are gain suppression, spatial hole burning, spontaneous emission at low power levels, and the relaxation oscillations at high modulation frequencies. The distortion caused by the intrinsic nonlinearities increases rapidly in the vicinity of the resonance frequency and usually limits the maximum bandwidth. The spontaneous emission only causes a significant distortion if a laser is biased near its threshold current. This is never the case in practical systems, and this cause of distortion will therefore not be given any further consideration in this chapter. For the same reason, we ignore thermal effects. They only have an influence on the dynamic behavior of a laser at modulation frequencies below 10 MHz.

In this chapter, we will mainly focus on the second- and third-order harmonic distortion. The higher order distortions become smaller with the order and less important if the optical modulation depth (OMD) is not too large. Moreover, the causes of all harmonic and intermodulation distortions are practically identical. The different causes of harmonic distortion will be treated separately first. However, due to the intrinsic nonlinear nature, the total distortion is not just the superposition of the distortions caused by the separate nonlinearities, and therefore possible combined effects are discussed as well. We end this chapter with a brief discussion of the composite second order (CSO) and composite triple beat (CTB) quantities, which are used to characterize analog systems, and with a brief introduction to the design of DFB lasers with low distortion. We only consider lasers with a stable single-mode behavior and leave the mode index behind in all formulas.

8.1 MEASURING THE HARMONIC DISTORTION

The measurement of the harmonic distortion is very similar to the measurement of the intensity modulation, except that one has to detect the intensity variations at a harmonic of the modulation frequency. The main obstacle in this measurement can be the distortion that is already present in the electrical signal generator or the distortion that is added by the photodetector [4] or the amplifier.

The distortion in an avalanche photodiode (APD) diode mainly occurs at high optical powers or in a state with large amplification, and it results from the sharp increase in amplification with increasing bias voltage. The distortion in a PIN diode mainly occurs at low bias voltages, where photovoltaic effects can be observed.

The harmonic distortion of the electrical signal generator is typically –30 dB, and the only way to get rid of its influence is by inserting a tunable bandpass filter after the signal generator [5]. Additional neutral density filters can be inserted before the photodiode to further suppress the distortion from this photodiode. To avoid problems with the distortion of the signal generator, the measurement of the harmonic distortion is often replaced by a measurement of the intermodulation distortion. The intermodulation distortion is related to the harmonic distortion, and its measurement is not affected by distortions in the signal generators. A typical setup for this measurement is shown in Figure 8.1.

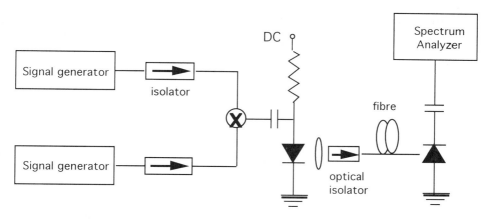

Figure 8.1 Measurement setup for the intermodulation distortion.

8.2 INFLUENCE OF THE RELAXATION OSCILLATIONS

At modulation frequencies in the vicinity of the resonance frequency, the harmonic distortion caused by the intrinsic nonlinearity is much larger than the harmonic distortion caused by spatial hole burning, leakage currents, or gain suppression. The analytical approximations for second- and third-order distortion, which can be derived from the rate equations, are then an accurate description, even for DFB lasers. These approximations have already been given in Chapter 2. The second- and third-order harmonic distortions caused by the intrinsic nonlinearity can be expressed as a function of the modulation pulsation Ω as [6,7]

$$\frac{S_2}{S_1} = \frac{j\Omega\left(2j\Omega + \dfrac{1}{\tau_d}\right)}{\left\{(2j\Omega + \xi G_{th}S_0)\left(2j\Omega + \dfrac{1}{\tau_d} + \dfrac{\partial G}{\partial N}S_0\right) + \left(\dfrac{\partial G}{\partial N} - \dfrac{\partial \gamma}{\partial N}\right)G_{th}S_0\right\}}\frac{m}{2}$$

$$\frac{S_3}{S_1} = \frac{1}{2}\frac{\left(3j\Omega + \dfrac{1}{\tau_d}\right)\left(3j\Omega \dfrac{S_2}{S_1}m - \dfrac{j\Omega}{2}m^2\right)}{\left\{(3j\Omega + \xi G_{th}S_0)\left(3j\Omega + \dfrac{1}{\tau_d} + \dfrac{\partial G}{\partial N}S_0\right) + \left(\dfrac{\partial G}{\partial N} - \dfrac{\partial \gamma}{\partial N}\right)G_{th}S_0\right\}} \quad (8.1)$$

with m the optical modulation depth, S_0 the average steady-state photon density, and S_i the component of the average photon density at frequency $i\Omega$. N is the average carrier density, G the modal gain, γ the modal loss, τ_d the differential carrier lifetime, and ξ the gain suppression coefficient. G_{th} is the threshold value of G. The effect of carrier transport can be taken into account by replacing $1/\tau_d$ in (8.1) using the transformation [8]

$$\frac{1}{\tau_d} \rightarrow \frac{1}{\tau_d} + \frac{j\omega\tau_s}{1+j\omega\tau_s}\frac{1}{\tau_e} \qquad (8.2)$$

with $\omega = 2\Omega$ for the second-order harmonic distortion and $\omega = 3\Omega$ for the third-order harmonic distortion. τ_e is the escape time from the quantum wells and τ_s is the transport time over the separate confinement layers. We will, however, neglect this effect further on. The distortions (8.1) are shown as a function of modulation frequency in Figures 8.2 and 8.3.

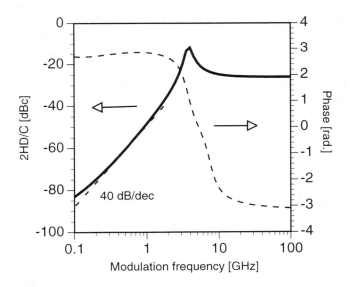

Figure 8.2 Second-order distortion caused by intrinsic nonlinearities.

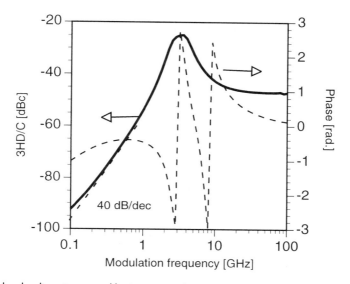

Figure 8.3 Third-order distortion caused by intrinsic nonlinearities.

It can be seen that 2HD/C and 3HD/C become constant beyond the resonance frequency. The limit values are given by

$$\frac{S_2}{S_1} = \frac{m}{4} \text{ and } \frac{S_3}{S_1} = \frac{m^2}{24} \qquad (8.3)$$

It is also possible to calculate a maximum, for example, for the second-order distortion:

$$\left|\frac{S_2}{S_1}\right|^2 = \frac{\Omega_r^4}{\theta^2(4\Omega_r^2 - \theta^2)} m \text{ for } \Omega^2 = \frac{1}{2} \frac{\Omega_r^4}{(2\Omega_r^2 - \theta^2)} \qquad (8.4)$$

with Ω_r the resonance pulsation ($=2\pi f_r$, with f_r the resonance frequency) and θ the damping [9,10]

$$\Omega_r^2 = (2\pi f_r)^2 = G_{th}\left(\frac{\partial G}{\partial N} - \frac{\partial \gamma}{\partial N}\right)S_0$$

$$\theta = \frac{1}{\tau_d} + \frac{\partial G}{\partial N}S_0 + G_{th}\xi S_0 \qquad (8.5)$$

The expressions (8.3) and (8.4) are of little importance, since the distortion near and beyond the resonance frequency is too large for analog applications in that frequency range.

As can be seen from Figures 8.2 and 8.3, the second- and third-order distortions increase dramatically above a few hundred megahertz. In CATV systems, where the harmonic distortions are usually required to be below −65 dBc, it is this rise of distortions that determines the maximum usable bandwidth. This maximum usable bandwidth is typically 1 to 2 GHz and still far below the resonance frequency. The distortion in the considered frequency range can be approximated as

$$\frac{S_2}{S_1} = \frac{j\Omega\left(2j\Omega + \frac{1}{\tau_d}\right)}{(2\pi f_r)^2}\frac{m}{2}$$

$$\frac{S_3}{S_1} = \frac{1}{2}\frac{j\Omega\left(3j\Omega + \frac{1}{\tau_d}\right)\left(3m\frac{S_2}{S_1} - \frac{m^2}{2}\right)}{(2\pi f_r)^2} \qquad (8.6)$$

and is thus mainly determined by the value of the resonance frequency and of the modulation frequency. Hence, the only way to reduce the distortion in this frequency range or to increase the useful bandwidth is to increase (optimize) the resonance frequency. Methods for achieving this have been discussed in the previous chapter.

8.3 INFLUENCE OF GAIN SUPPRESSION

The gain suppression has no influence on the actual value of the gain, but it results in an increase of the carrier density (required to obtain the threshold gain) with increasing power. The distortion in the AM response caused by this gain suppression therefore has its origin in the increase of the internal loss with increasing power and the increase of the spontaneous carrier recombination with power. We already remark that a similar effect can be expected to result from the presence of leakage currents. Both effects will result in a sublinear P/I-curve, and the second-order distortion will have a phase π at low modulation frequencies.

After manipulation of the rate equations, one finds the following expressions for the second- and third-order distortions at low modulation frequencies [11–13]:

$$\frac{S_2}{S_1} = -\left\{ \frac{\frac{\partial \gamma}{\partial N}}{\left(\frac{\partial G}{\partial N} - \frac{\partial \gamma}{\partial N}\right)} \xi S_0 + \frac{\frac{\partial G}{\partial N}}{\left(\frac{\partial G}{\partial N} - \frac{\partial \gamma}{\partial N}\right)^2} \xi^2 \frac{S_0}{\tau_d} \right\} \frac{m}{2}$$

$$\frac{S_3}{S_1} = -\left\{ \frac{\frac{\partial \gamma}{\partial N}\frac{\partial G}{\partial N}}{\left(\frac{\partial G}{\partial N} - \frac{\partial \gamma}{\partial N}\right)^2} (\xi S_0)^2 + \frac{\left(\frac{\partial G}{\partial N}\right)^2}{\left(\frac{\partial G}{\partial N} - \frac{\partial \gamma}{\partial N}\right)^3} \xi^3 \frac{S_0^2}{\tau_d} \right\} \frac{m^2}{4} + 2\left(\frac{S_2}{S_1}\right)^2 \quad (8.7)$$

The carrier density dependence of the differential carrier lifetime τ_d has been neglected in the derivation of this formula.

The influence of gain suppression through the carrier density dependence of the absorption often dominates the second-order distortion in bulk lasers (because of their "small" differential gain) and MQW lasers (because of the large value of the differential absorption). The second-order harmonic distortion caused by gain suppression is shown as a function of bias power in Figure 8.4 for a 600-μm-long, cleaved MQW DFB laser with the following material parameters: $dg/dN = 7 \times 10^{-16}$ cm^2, $d\alpha/dN = 1.6 \times 10^{-16}$ cm^2, and $\xi = 4 \times 10^{-17}$ cm^3 [14,15]. The increase of the distortion with κL that can be observed in Figure 8.4 is due to the dependence of the distortion on the average photon density and the increase of average photon density with increasing κL.

The different terms in (8.7) can be estimated from an analytical calculation, for example, assuming a uniform power level of 10 mW inside the cavity and an optical modulation depth of 0.35. One finds that the first term in the first expression of (8.7) gives a second-order distortion of about −54 dB. The second term gives, for the same parameters, a second-order distortion of about −81 dB. The first and second terms of the second expression of (8.7) give a third-order distortion of about −95 dB and −122 dB, respectively, under the same conditions. This third-order distortion is, however, usually dominated by the distortion caused by spatial hole burning.

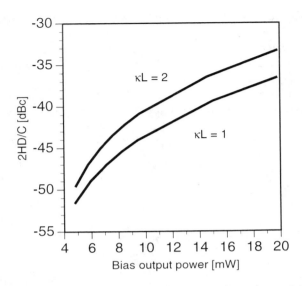

Figure 8.4 Second-order distortion caused by the carrier density dependence of the absorption for a cleaved, 600-μm-long MQW DFB laser.

If gain suppression is the main cause of harmonic distortion, a significant reduction of the second-order distortion (e.g., 6 dB or more) is usually possible by decreasing the average photon density S_0, for example, by applying an AR coating on one or both facets or by using a different active layer material with more favorable material constants. Another possibility is decreasing the active layer dimensions (especially the thickness). Since the confinement factor is proportional to the squared thickness for a small enough thickness, one decreases the average photon density S_0 corresponding to a certain output power, as well as the gain compression ξ (see Section 2.5.4) with this measure.

We will assume further on that the carrier density dependence of the absorption can be ignored, as is the case for strained-layer multi-quantum-well (SL-MQW) lasers.

8.4 INFLUENCE OF SPATIAL HOLE BURNING

The impact of spatial hole burning on the low-frequency harmonic distortion is twofold:

- Via a change in the threshold gain and the average threshold carrier density;

- Via a change of the relationship between output power and average photon density.

To illustrate the last effect in more detail, we consider a Fabry-Perot laser with facet reflectivities R_1 and R_2. The optical power is also nonuniform in this case (and so are, therefore, carrier density and gain). The power of forward and backward propagating waves can be written as

$$S^+(z) = S^+(0) \exp\left[\int_0^z (\Gamma g - \alpha_{int}) \, dz'\right]$$

$$S^-(z) = S^-(0) \exp\left[-\int_0^z (\Gamma g - \alpha_{int}) \, dz'\right] \quad (8.8)$$

Taking into account that $S^+(0) = R_1 S^-(0)$ and $S^-(L) = R_2 S^+(L)$, one still finds the threshold condition

$$G(N_{av}, \lambda) = v_g\left(\alpha_{int} + \frac{1}{2L} \ln \frac{1}{R_1 R_2}\right) \quad (8.9)$$

That is, the average carrier density (or threshold gain) is independent of the power level. On the other hand, it has been shown in Chapter 3 that a nonuniformity of the optical power $\Delta S(z)$ results in a nonuniformity of the carrier density and changes the average stimulated emission rate $G(N(z), \lambda)S(z)$ into

$$\frac{1}{L}\int_0^L G(N(z), \lambda) S(z) \, dz = G(N_{av}, \lambda) S_{av}(1 - \xi_{spat} S_{av})$$

with

$$\xi_{spat} = \frac{\dfrac{\partial G}{\partial N}}{\dfrac{1}{\tau_d} + \dfrac{\partial G}{\partial N} S_{av}} \frac{1}{L}\int_0^L \left[\frac{\Delta S(z)}{S_{av}}\right]^2 dz \quad (8.10)$$

It is the average stimulated emission rate that has to equal the total loss, so that (8.9) and (8.10) are contradictory unless the facet loss γ_{fac} depends on the optical power.

The problem can be solved by calculating the average photon density from (8.8), using the approximation

$$\exp\left[\int_0^z \Gamma \frac{\partial g}{\partial N} \Delta N(z')dz'\right] = 1 + \int_0^z \Gamma \frac{\partial g}{\partial N} \Delta N(z')dz' \qquad (8.11)$$

After some calculations, one indeed finds that the facet loss γ_{fac} also depends on the average photon density

$$\gamma_{fac} = G(N_{av}, \lambda)(1 - \xi_{spat}S_{av}) - \alpha_{int}v_g \qquad (8.12)$$

Hence, the relationship between output power and average photon density (or average power inside the cavity) is nonlinear.

Taking the nonlinearity (8.12) into account results in the following expressions for the harmonic distortion caused by spatial hole burning in Fabry-Perot lasers

$$\frac{P_2}{P_1} = \frac{\alpha_{int}L\xi_{spat}S_0}{\ln(R_1R_2)} m$$

$$\frac{P_3}{P_1} = \frac{\alpha_{int}L\xi_{spat}^2 S_0^2}{\ln(R_1R_2)} m^2 \qquad (8.13)$$

That is, the distortion (in the output power and not in the photon density) is proportional to the internal loss. The second-order distortion is illustrated in Figure 8.5 for a Fabry-Perot laser. The gain suppression ξ_{spat} depends greatly on the facet reflectivities for such a laser, but is approximately 0.5/W for a 300-μm-long cleaved laser [11].

A similar effect is also present in DFB lasers. The gain suppression ξ_{spat} is still given by (8.10) for such lasers, but the power dependence of the facet loss becomes more complex. For small modulation frequencies, one can still write the carrier density rate equation as

$$\frac{I}{qV_{act}} = AN + BN^2 + CN^3 + \alpha_{int}v_gS + \frac{\Gamma P_{out}}{h\nu V_{act}} \qquad (8.14)$$

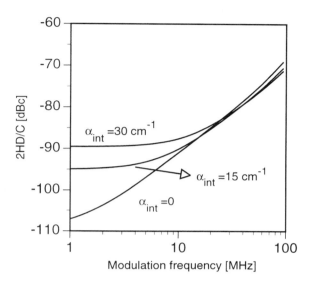

Figure 8.5 Second-order distortion in a Fabry-Perot laser caused by spatial hole burning. The bias power is 1 mW and m = 0.2.

with P_{out} the total output power from both facets. If the variations in the threshold current (or in the spontaneous carrier recombination rate) are neglected, the following distortion in the output power results

$$\frac{P_{out,2}}{P_{out,1}} = -\frac{\alpha_{int} v_g}{\gamma_{fac}} \frac{S_2}{S_1} \quad \text{and} \quad \frac{P_{out,3}}{P_{out,1}} = -\frac{\alpha_{int} v_g}{\gamma_{fac}} \frac{S_3}{S_1} \quad (8.15)$$

From (8.15) one can expect that the distortion in the output power will decrease with decreasing internal loss, although it has to be taken into account that the distortion in the photon density, S_2/S_1 or S_3/S_1, depends on the internal loss as well. Moreover, (8.15) gives the distortion in the total output power from both facets. It will only equal the distortion in the output power from a single facet if the output powers from both facets remain in a constant ratio independent of the bias level. The influence of the second-order variations in the carrier recombination is usually relatively small.

The distortion in the photon density caused by spatial hole burning is not easily described analytically. We therefore first illustrate some aspects of the spatial hole burning–induced distortion for λ/4-shifted DFB lasers with both facets perfectly AR-coated. The effect of spatial hole burning in such lasers is notably different for small and large κL values, and this is also reflected in the harmonic distortion. In particular, large κL values give rise to a power profile that is concentrated in the center of the laser, while small κL values give rise to a power profile that is concentrated near the facets. The difference in distortion resulting from that difference in spatial hole burning can be seen from Figures 8.6 and 8.7, which show the second-order distortion at 1-mW bias output power as a function of modulation frequency for a 400-μm-long, AR-coated λ/4-shifted laser with κL = 1 and κL = 2, respectively. It can be seen that the second-order distortion has a phase π for κL = 1 and a phase 0 for κL = 2.

From a more extensive numerical study, it can be concluded that a power profile concentrated in the middle of the laser cavity generally results in a phase 0 for the second-order distortion, while a power profile concentrated near the facets generally results in a phase π. This rule holds for all laser structures, provided that the internal loss is not too small. In our calculations, we assumed an internal loss of 40 cm^{-1}. The third-order distortion of the λ/4-shifted lasers has a phase π for both values of κL. It is depicted in Figure 8.8 for κL = 1 and a bias output power of 1 mW. In principle, both a phase 0 and a phase π should be possible here as well.

The contribution of spatial hole burning to the distortion in the photon density (i.e., S_2/S_1 and S_3/S_1) obviously depends on the nonuniformity of the optical power inside the cavity and thus on the coupling strength and the facet reflectivities [16–19]. The optical power is usually more uniform in lasers with high facet reflectivities (e.g.,

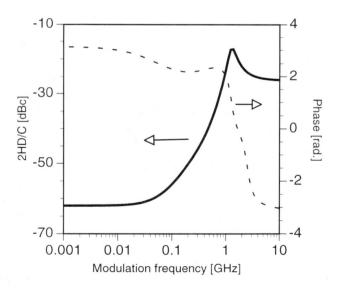

Figure 8.6 Second-order distortion of a λ/4-shifted DFB laser with κL = 1. The bias power is 1 mW and m = 0.2.

Harmonic and Intermodulation Distortion in DFB Laser Diodes

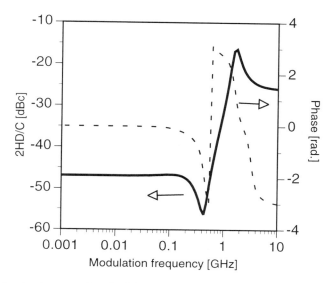

Figure 8.7 Second-order distortion of a $\lambda/4$-shifted laser with $\kappa L = 2$. The bias power is 1 mW and $m = 0.2$.

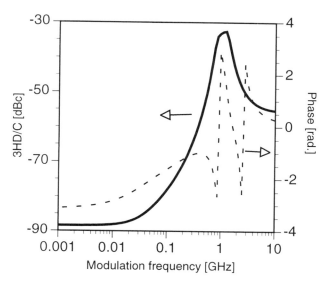

Figure 8.8 Third-order distortion of a $\lambda/4$-shifted laser with $\kappa L = 1$. The bias power is 1 mW and $m = 0.2$.

better for cleaved lasers than for AR-coated lasers such as the λ/4-shifted laser) and small κL constant. However, the single-mode yield also decreases as the facet reflectivities become larger and the κL value becomes smaller. A compromise, resulting in an acceptable single-mode yield (e.g., 30%) and in a relatively small distortion, is the use of lasers with cleaved facets and with a relatively small κL value (e.g., between 0.5 and 1) [20–22].

A comparison between the second- and third-order distortions of lasers with different κL values is given in Figures 8.9 and 8.10. Figure 8.11 shows the second-order distortion for the same lasers but with one facet AR-coated and having a facet reflectivity of 5%. Similar results can be obtained for the third-order distortion. As expected, the distortion of the AR-coated lasers is considerably larger than that of cleaved lasers. It also increases with bias power over a long range. This behavior can be observed for most lasers with high threshold gain (e.g., the λ/4-shifted lasers with both facets AR-coated). As will become clear further on, it is caused by the relatively small differential carrier lifetime, the relative small differential gain, and the relatively small ratio of photon density and output power in lasers with high threshold gain.

As mentioned in Chapter 6, there is indeed a strong dependence of the spatial hole burning on the threshold gain, the differential gain, and the carrier lifetime.

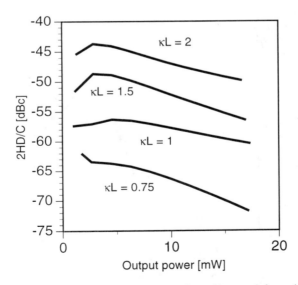

Figure 8.9 Static second-order distortion of a 600-μm-long cleaved laser with facet phases $\phi_1 = 0$, $\phi_2 = \pi/2$ for different κL values and for an OMD of 35%. Gain suppression not included.

Harmonic and Intermodulation Distortion in DFB Laser Diodes

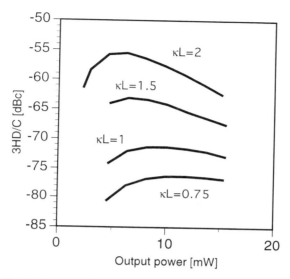

Figure 8.10 Static third-order distortion of a 600-μm-long cleaved laser with facet phases $\phi_1 = 0$, $\phi_2 = \pi/2$ for different κL values and for an OMD of 35%.

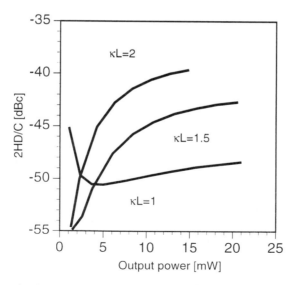

Figure 8.11 Static second-order distortion of a 600-μm-long laser with one AR-coated and one cleaved facet and facet phases $\phi_1 = 0$, $\phi_2 = \pi/2$ for different κL values and for an OMD of 35%.

The nonuniformity of the carrier density caused by the nonuniformity of the optical power can be estimated from a rate equation for the local carrier density $N(z)$ using local photon densities $S(z)$ (number of photons per unit length). Such an expression was derived in Section 6.3. With the notation used here, the nonuniform carrier density $\Delta N(z) = N(z) - N_0$ as a function of the nonuniform photon density $\Delta S(z) = S(z) - S_0$ is

$$\Delta N(z) = - \frac{G_{th} \Delta S(z)}{\dfrac{1}{\tau_d} + \dfrac{\partial G}{\partial N} S_0} \tag{8.16}$$

with S_0 and N_0 being the average photon and carrier density, respectively. It can also be assumed that the longitudinal power profile is independent of the bias level so that $\Delta S(z) = S_0 f(z)$ (i.e., $f \equiv \Delta S(z)/S_0$ is independent of S_0, as is the case in many lasers). Equation (8.16) can then be regarded as an explicit expression of the total variation in the carrier density at a particular bias level. Derivation of this expression with respect to S_0 then gives the following expression for the change in the axial variation of the carrier density with changing bias level

$$\delta(\Delta N(z)) = - \frac{G(N_0) f(z) \delta S_0 \tau_d}{\left(1 + \dfrac{\partial G}{\partial N} S_0 \tau_d\right)^2} \tag{8.17}$$

The variation of the average photon density $\delta S_0 = m S_0$ increases linearly with the bias output power if a constant modulation depth m is considered.

At very low bias levels, the bias dependence of the denominators can be neglected and (8.17) shows that the nonuniformity of the carrier density increases with bias level and becomes weaker with decreasing threshold gain. The carrier density variations are in this case also proportional to the small-signal carrier lifetime τ_d. Since τ_d increases with decreasing carrier density, this weakens the decrease of ΔN with decreasing threshold gain somewhat. Obviously, the contribution of spatial hole burning not only depends on the magnitude of the carrier density variations, but also on the influence of the carrier density variations on refractive index and distributed feedback loss. The influence of carrier density variations on the feedback loss can be quite different from one structure to another and depends, for example, on the length and the coupling strength. At very high photon densities, the second term in the denominator dominates. The small-signal spatial hole burning and the harmonic distortion decrease with bias power in this power range. This decrease of the distortion is obviously faster for larger differential gain and larger differential carrier lifetime.

The influence of the differential gain can be illustrated by considering different values for the confinement factor or, equivalently, different values for the active layer thickness or the number of quantum wells. Numerical results on second- and third-order distortions showing the influence of the active layer thickness on the spatial hole burning–induced distortion are depicted in Figures 8.12 and 8.13. The influence of the internal loss on the second-order distortion is illustrated in Figure 8.14 for a cleaved DFB laser with $\kappa L = 1$.

It must finally be noted that the distortion caused by spatial hole burning generally depends on the quantities that also affect the influence of spatial hole burning on the side-mode suppression. The influence of the series resistance on the harmonic distortion can therefore also be estimated from (6.9). Since the cutoff frequency of spatial hole burning is normally smaller than the resonance frequency, it has no influence on the damping of the relaxation oscillations. This damping is practically identical to the value that can be derived from (2.41).

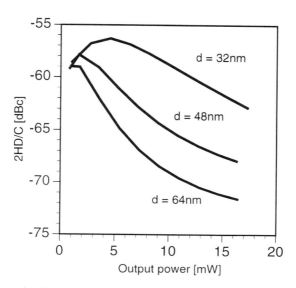

Figure 8.12 Static second-order distortion of a 600-μm-long cleaved laser with $\kappa L = 1$ and facet phases $\phi_1 = 0$, $\phi_2 = \pi/2$ for different d-values and for an OMD of 35%.

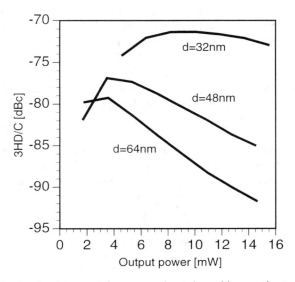

Figure 8.13 Static third-order distortion of a 600-μm-long cleaved laser with $\kappa L = 1$ and facet phases $\phi_1 = 0$, $\phi_2 = \pi/2$ for different d-values and for an OMD of 35%.

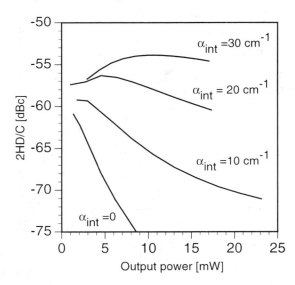

Figure 8.14 Influence of the internal absorption α_{int} on the spatial hole burning–induced second-order distortion for a 600-μm-long cleaved laser with $\kappa L = 1$ and facet phases $\phi_1 = 0$, $\phi_2 = \pi/2$.

8.5 INFLUENCE OF LEAKAGE CURRENTS

The influence of leakage current on the harmonic and intermodulation distortions has not been studied intensively. It is, however, independent of the longitudinal structure and identical for Fabry-Perot and DFB lasers. The effect can generally be described by replacing the current I in the carrier rate equation with a nonlinear function of the current $f(I)$; that is,

$$\frac{dN}{dt} = \frac{f(I)}{qV_{\text{act}}} - AN - BN^2 - CN^3 - G(N, S)S \qquad (8.18)$$

The function $f(I)$ will, however, generally depend on the modulation frequency (or, in the time domain, include derivatives of the current with time).

A useful treatment of the distortion caused by current leakage has been given by Lin et al. [23,24]. The approach is based on the equivalent circuit shown in Figure 8.15, in which the leakage current is assumed to be mainly due to the presence of a p-n homojunction diode. For this simple circuit, one can express the injected current as the sum of the leakage current I_l flowing through the homojunction and the current I_a flowing through the active region. In the derivation of the distortion caused by the leakage, it is assumed that the voltage V_a across the laser diode saturates, that is, $V_a = V_g$, with V_g the bandgap voltage. From Kirchhoff's law, it also follows that the voltage V_l across the homojunction diode is given by

$$V_l = V_g + R_a I_a - R_l I_l \qquad (8.19)$$

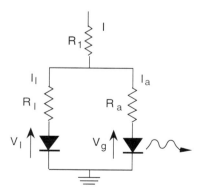

Figure 8.15 Equivalent circuit of a laser diode used to calculate the distortion caused by current leakage. (*After:* [24], © 1990 IEEE.)

with R_a and R_l the series resistances (apart from R) of the laser diode and of the leakage path. From the solution of the time-dependent diffusion equation, it can be shown that the response of the leakage diode to a current modulation of $I = I_0[1 + m\exp(j\Omega t)]$ is given by

$$I_l = I_s\left[\exp\left(\frac{qV_l}{nkT}\right) - 1\right]$$

with

$$I_s = I_{s0}\{1 + m\gamma_0 e^{j(\Omega t + \theta/2)}\}, \; \theta = tg^{-1}(\Omega\tau)$$

$$I_{s0} = \frac{qAD}{L} n_{p0} \tag{8.20}$$

with τ, D, L, and n_{p0} the carrier lifetime, diffusion coefficient, diffusion length, and equilibrium carrier density in the current blocking layer, respectively. A is the cross section of the leakage path.

The equations of (8.20) can be used to derive the leakage current as a function of the injected current. One finds the implicit relation

$$I_l = I_s[\beta \exp(aI - bI_l) - 1]$$

$$a = \frac{qR_a}{nkT}, \; b = \frac{q(R_a + R_l)}{nkT}, \; \beta = \exp\left(\frac{qV_g}{nkT}\right) \tag{8.21}$$

whose series expansion allows the calculation of the components of the current through the active layer $(I - I_l)$ at harmonics of the modulation frequency.

The effect of this current leakage on the second-harmonic distortion is illustrated in Figure 8.16 for a 10% current modulation and for laser emitting at 1.3 μm. From this figure it can be concluded that the distortion caused by current leakage is strongly bias-dependent and actually goes through a sharp minimum as a function of bias current. The minimum becomes less sharp, however, as the modulation frequency increases or as the carrier lifetime τ increases. For low modulation frequencies, it occurs when the potential drop over both resistances R_a and R_l is equal. The potential V_l over the leakage diode then remains clamped to a value V_g that is constant and determined by the threshold carrier density in the active layer. With an appropriate

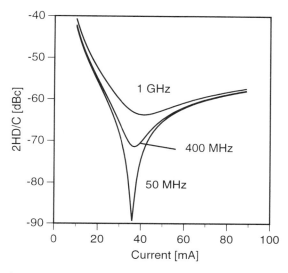

Figure 8.16 Harmonic distortion caused by current leakage as a function of bias current and at different frequencies for m = 0.1. (*After:* [24], © 1990 IEEE.)

design of resistances R_a and R_l, one can theoretically have this minimum occurring at any desired bias current.

The current leakage and its effect on the harmonic distortion generally depend on the emission wavelength as well. This dependence is pronounced in the dependence of β on V_g. For example, for emission at 1.3 μm and with $n = 2$ and $T = 300K$, one has β = 124, while for λ = 1.55 μm one has β = 15. The reduced β-value in the case of 1.55 μm will be reflected in a reduction of the distortion caused by current leakage with a factor of ±10. From (8.21) it can be seen that the distortion caused by leakage will become lower if the resistance R_a decreases or if the resistance R_l increases. Indeed, the leakage current becomes less dependent on the injected current if R_a or a decreases (for $a = 0$, a constant leakage current results). If R_l or b increases, the leakage current will become smaller in amplitude (and zero for $b = \infty$).

The approach given before is, strictly speaking, only valid for laser structures such as capped-mesa buried heterostructure (CMBH) or similar structures, which can be represented by the equivalent circuit of Figure 8.15. A general approach, valid also for laser structures that cannot be represented by the circuit of Figure 8.15, would be to carefully inspect the expressions given in Chapter 5 for lateral leakage and leakage over the heterojunctions and to perform a small-signal analysis of these expressions to determine their second- and third-order components. This would, however, lead to complex calculations that require numerical implementation. Such an approach has not been reported yet.

8.6 DIPS IN THE BIAS AND FREQUENCY DEPENDENCE OF THE DISTORTION

As has been seen before, the total harmonic distortion in the output of a laser diode is not the sum of the distortions caused by the individual nonlinear effects. When one would be able to express the total distortion analytically, one would undoubtedly find terms that cannot be separated into different nonlinearities. We illustrate this here for the case in which only the gain suppression and the intrinsic nonlinearities are included. The second-order distortion in this case can be expressed as

$$\frac{S_2}{S_1} = \frac{\left(2j\Omega + \dfrac{1}{\tau_d}\right)(j\Omega - \xi^2 S_0^2 G_{\text{th}})}{\left\{(2j\Omega + \xi G_{\text{th}} S_0)\left(2j\Omega + \dfrac{1}{\tau_d} + \dfrac{\partial G}{\partial N} S_0\right) + \left(\dfrac{\partial G}{\partial N} - \dfrac{\partial \gamma}{\partial N}\right) G_{\text{th}} S_0\right\}} \frac{m}{2} \quad (8.22)$$

The product $2j\Omega\xi^2(S_0)^2 G_{\text{th}}$ in the denominator is a term having its origin in the combination of gain suppression and intrinsic nonlinearities. Similarly, nonlinear combinations of gain suppression and spatial hole burning will be present.

Since the distortion caused by spatial hole burning can have a phase 0, while the distortion caused by gain suppression always has a phase π, both nonlinearities can cancel each other out to some extent. This canceling occurs only when both contributions have approximately the same strength and, because of the specific bias dependence of both contributions, only in a certain range of bias powers. Indeed, spatial hole burning becomes weaker with increasing bias, while the gain suppression effect increases with bias level. The canceling is illustrated in Figure 8.17 for a cleaved DFB laser with $\kappa L = 0.75$ and $L = 600$ µm. At low bias levels, the gain suppression contribution is very weak and spatial hole burning dominates. As the bias level increases, however, the distortion caused by spatial hole burning (with phase 0) decreases and the contribution from gain suppression (with phase π) increases. From a certain bias level on, both nonlinearities start to cancel each other out and the total low-frequency distortion becomes a lot smaller than the distortion caused by the individual nonlinearities.

A similar effect can originate from the interference between the contributions from spatial hole burning and the intrinsic nonlinearity. The distortion caused by the last nonlinearity has a phase π below the resonance frequency and increases with modulation frequency in that frequency range. Therefore, if the spatial hole

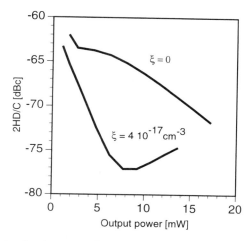

Figure 8.17 Static second-order distortion of a 600-µm-long cleaved laser with facet phases $\phi_1 = 0$, $\phi_2 = \pi/2$ with and without the inclusion of gain suppression and for an OMD of 35%.

burning–induced distortion has a phase 0, both spatial hole burning and the intrinsic nonlinearity will cancel each other in a certain frequency range. The effect can be seen in Figure 8.7 for a λ/4-shifted DFB laser with κL = 2, but it also occurs for most lasers with cleaved facets. This particular interference effect has the advantage that the distortion remains low up to a higher frequency range (i.e., it is useful to obtain a larger bandwidth).

8.7 RELATION WITH CSO AND CTB

A system design usually imposes no restriction on the harmonic distortion, but rather on the system-dependent quantities such as the CSO and the CTB. For mutual harmonic carriers, the CSO can be defined as the ratio of the peak power in the carrier to the peak power in the second-order intermodulation tone, and the CTB can be defined as the ratio of the peak power in the carrier to the peak power in the third-order intermodulation tone. The relationship between the system-dependent quantities CSO and CTB and the laser-dependent quantities, second- and third-order harmonic distortion, will only be addressed briefly here. The reader can refer to the literature [25,26] for a more detailed account.

We first show how the intermodulation distortion is related to the harmonic distortion. We assume for the sake of simplicity that only three carriers (with frequencies $f_1, f_2,$ and f_3 and with an identical OMD) are used and that the current injection is therefore given by

$$I = I_{\text{th}} + I_0\left\{1 + \frac{m}{3}\left[\cos(\Omega_1 t) + \cos(\Omega_2 t) + \cos(\Omega_3 t)\right]\right\} \quad (8.23)$$

m is the total modulation depth. We also assume that the harmonic distortion is nearly flat in the frequency region that contains f_1, f_2, and f_3. The optical output power can then be approximated as

$$P_{\text{out}} = \eta_1(I - I_{\text{th}}) + \eta_2(I - I_{\text{th}})^2 + \eta_3(I - I_{\text{th}})^3 \quad (8.24)$$

After substituting (8.23) into (8.24), one finds that the light intensity not only consists of components with frequency f_i (linear response), $2f_i$ (second-order harmonic distortion), and $3f_i$ (third-order harmonic distortion), but also components at the frequencies $f_i \pm f_j$ and $f_i \pm f_j \pm f_k$. Moreover, it can easily be checked that the amplitude of the components at the frequencies $f_i \pm f_j$ ($i \neq j$) is a factor 2 (6 dB) larger than the amplitude of the components at the frequencies $2f_i$. The amplitude of the components at the frequencies $f_i \pm f_j \pm f_k$ ($i \neq j \neq k$) is a factor 6 (15.5 dB) larger than the amplitude of the components at frequencies $3f_i$ and a factor 2 (6 dB) larger than the amplitude of the components at frequencies $2f_i \pm f_j$ ($i \neq j$). The intermodulation distortion can simply be derived from the harmonic distortion with the help of these relations.

Simple relationships between harmonic distortion and intermodulation distortion also exist for modulation and intermodulation frequencies, where the intrinsic nonlinearity gives a dominating contribution to the distortion. The expressions for the third-order intermodulation distortion can be derived using small-signal approximations of the rate equations, but they become very lengthy, and we will therefore only give the expression for the third-order intermodulation product at frequencies $2f_i \pm f_j$:

$$S_2(\Omega_3 = \Omega_1 \pm \Omega_2) = \frac{1}{2} \frac{j\Omega_3\left(j\Omega_3 + \dfrac{1}{\tau_d}\right)S_1(\Omega_1)S_1(\Omega_2)}{\left\{(j\Omega_3 + \xi G_{\text{th}}S_0)\left(j\Omega_3 + \dfrac{1}{\tau_d} + \dfrac{\partial G}{\partial N}S_0\right) + \dfrac{\partial G}{\partial N}G_{\text{th}}S_0\right\}S_0}$$

$$S_3(\Omega_3 = 2\Omega_1 - \Omega_2) = \frac{1}{2} \frac{\left(j\Omega_3 + \dfrac{1}{\tau_d}\right)N}{\left\{(j\Omega_3 + \xi G_{\text{th}}S_0)\left(j\Omega_3 + \dfrac{1}{\tau_d} + \dfrac{\partial G}{\partial N}S_0\right) + \dfrac{\partial G}{\partial N}G_{\text{th}}S_0\right\}S_0}$$

$$N = j\left[\Omega_1 S_1(\Omega_1)S_2(\Omega_1 - \Omega_2) + \Omega_2 S_1(\Omega_2)S_2(2\Omega_1) - \frac{\Omega_1 - \Omega_2}{2}S_1^2(\Omega_1)S_1(\Omega_2)\right] \quad (8.25)$$

The CSO and the CTB for a given carrier can then be obtained as the sum of all possible intermodulation products with a frequency in this channel. In the application of SCM in fiber-optic microcellular radio communication, a carrier with frequency 1.9 GHz is multiplexed by a number of channels with considerably smaller frequency. Only the third-order intermodulation products are of importance in this case. In the application of SCM for CATV, a large amount of carriers are located between 50 MHz and 1.7 GHz. Both the second- and third-order intermodulation products must then be limited.

When a "large" number of channels or subcarriers are used in the system, one must also take into account possible clipping effects. These occur when the optical signal resulting from all modulated carriers exceeds the bias power; that is, when due to the modulation of all carriers, the injected current decreases below the threshold current (see Figure 8.18). The optically transmitted power and the current at the receiver is reset to zero at all these instants. A simple treatment of this "clipping" phenomenon has been given by Saleh [27]. In his derivation, Saleh assumed that the receiver current is proportional to the received optical power:

$$I(t) = \rho P(t) \tag{8.26}$$

and that the optical power without clipping can be expressed as

$$P(t) = P_0 \left[1 + \sum_{i=1}^{N} m_i \cos(\Omega_i t + \varphi_i) \right] \tag{8.27}$$

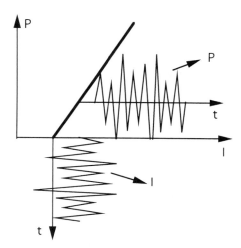

Figure 8.18 Analog system with clipping-induced distortion.

The receiver current I can then be accurately modeled as a Gaussian random process with a mean value of I_0 ($=\rho P_0$) and a variance of $(\sigma_I)^2 = (I_0)^2 Nm^2/2$ if the number of channels N is sufficiently large (e.g., $N > 10$). The total mean-square value of the clipped portion of $I(t)$ (i.e., the portion that is below zero) becomes

$$\langle I_{NLD}^2 \rangle = \frac{1}{\sqrt{2\pi\sigma_I^2}} \int_{-\infty}^{0} I^2 \exp[-(I - I_0)^2/2\sigma_I^2] \, dI = \sqrt{\frac{2}{\pi}} I_0^2 \mu^5 \exp\left(-\frac{1}{2\mu^2}\right)$$

$$\mu = \sqrt{\frac{Nm^2}{2}} \qquad (8.28)$$

with μ the total RMS modulation index. The mean-square value of the current in each channel is $(I_0)^2 m^2/2$, and the distortion-to-carrier ratio per channel can be expressed as

$$\text{NLD/C} = \sqrt{\frac{2}{\pi}} \mu^3 \exp\left(-\frac{1}{2\mu^2}\right) \qquad (8.29)$$

This function gives an overestimation, since it assumes that all the clipped power is uniformly distributed over the different channels. It was also derived under the assumption that the modulation depth μ is small.

A more accurate expression for the CSO and CTB caused by clipping has been derived in [28] and is given by

$$\text{CSO} = \frac{K_2}{16\pi N} \exp\left(-\frac{2}{\mu^2}\right)\left[1 + \text{erf}\left(\frac{1}{\mu}\right)\right]$$

$$\text{CTB} = \frac{K_3}{32\pi N^2} \frac{1}{\mu^2} \exp\left(-\frac{2}{\mu^2}\right)\left[1 + \text{erf}\left(\frac{1}{\mu}\right)\right] \qquad (8.30)$$

K_2 and K_3 are the number of intermodulation products that fall within the considered channel and erf(.) is the error function. The CSO is plotted in Figure 8.19 for channel 3 of a 42-channel NTSC system as a function of μ. Distorting this channel are 120 intermodulation products. It can be seen that the clipping distortion is relatively small and negligible as long as the modulation depth remains below ± 0.4.

From (8.27), it follows that the clipping distortion can be completely prevented if the modulation index m of the channels is limited to $1/N$. Such a restriction is, however, too conservative and it is more advantageous to accept some clipping and use a modulation index m a little beyond $1/N$. There is distortion and noise anyway.

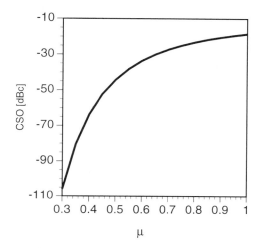

Figure 8.19 Clipping distortion versus the rms modulation index for channel 3 of a 42-channel NTSC system. (*After:* [28], © 1992 IEEE.)

8.8 DESIGNING HIGHLY LINEAR DFB LASERS

Using the theory outlined before, it is possible to design DFB lasers with small harmonic distortion over a large frequency range. Several such linear lasers from different manufacturers have been reported in the literature [17,18,29,30]. Practically all these devices possess an SL-MQW active layer. The small carrier density dependence of the internal loss in this material allows us to ignore the contribution from gain suppression up to a large output power. If gain suppression has influence, it is rather positive. It cancels the contribution from spatial hole burning to some extent at high power levels.

To obtain a low distortion over a large frequency range, one could try to make use of the occurrence of dips in the frequency dependence of the distortion. This extends the useful bandwidth, but only occurs if the static (i.e., spatial hole burning–induced) distortion has an appropriate phase (i.e., 0 and π for the second- and third-order distortion, respectively). The phase 0 (π) for the second- (third-) order distortion is most probably obtained for high κL values and rather high facet reflection coefficients.

It is nevertheless beneficial to have the individual causes of distortion, spatial hole burning, and intrinsic nonlinearities as low as possible. A weak intrinsic nonlinearity is obtained for a high value of the resonance frequency. Possible measures that can be taken to achieve this include an increase of the confinement factor (or the active layer dimensions), a decrease of the facet loss per unit length, a decrease of the transparency carrier density (e.g., achieved by doping the active layer), and an increase of the logarithmic gain coefficient g_0 (see (2.27)). Furthermore, there exists an optimum length given by [31]

$$L_{opt} = \frac{f(R_1, R_2, \kappa L)}{\Gamma g_0 - \alpha_{int}}, f = 2\alpha_{fac} L \qquad (8.31)$$

with f the normalized facet loss (function of the facet reflectivities $R_{1,2}$ and the normalized coupling coefficient κL). The influence of the laser length L on the harmonic distortion is illustrated in Figure 8.20 for a cleaved DFB laser with $\kappa L = 1$ and facet phases $\phi_1 = \pi/4$, $\phi_2 = 7\pi/4$, for $\Gamma = 0.1035$, $\alpha_{int} = 20$ cm^{-1}, $g_0 = 2,350$ cm^{-1}, and $N_0 = 1.7 \times 10^{18}$ cm^{-3}. By reducing the length from 600 to 200 µm, over 20-dB improvement in second-order distortion is obtained at a modulation frequency of 450 MHz and a bias power of 5 mW. Note, however, that the optimum length depends greatly on the confinement factor and on g_0 (which is usually a lot larger than α_{int}). A decrease of the facet loss obtained through an increase of κL usually increases the spatial hole burning–induced distortion (see 8.4) and is therefore not recommended as a measure for resonance frequency enhancement. An increase of the facet reflectivities, on the other hand, can be acceptable and often implies a reduction of spatial hole burning. A serious drawback of this measure is the reduction of the single-mode yield.

The distortion caused by spatial hole burning and its reduction have been discussed before as well. Both second- and third-order distortions decrease with decreasing kL and with decreasing internal loss. However, this contribution depends on the

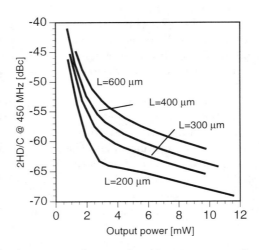

Figure 8.20 Second-order distortion as a function of total bias output power for a cleaved DFB laser with $\kappa L = 1$ and for different cavity lengths. $m = 0.35$, modulation frequency = 450 MHz.

Harmonic and Intermodulation Distortion in DFB Laser Diodes

confinement factor and on the cavity length. The influence of the confinement factor has been described in Section 8.4. The influence of the cavity length is illustrated in Figure 8.21 using the same parameters and considering a modulation frequency of 50 MHz. As was demonstrated clearly in Figures 8.9 and 8.10, the distortion resulting from spatial hole burning decreases substantially with decreasing κL. From a certain bias power on, it also decreases with this bias power. Hence, using a high bias power (or a high average photon density) is beneficial for both the low- and high-frequency distortion. There are nevertheless causes of distortion such as gain suppression and current leakage that can increase with bias level. To be able to limit the bias power, a small κL value is preferred.

Although the distortion decreases monotonically with decreasing κL, so does the single-mode yield and a compromise has to be found. Fortunately, the levels of CSO and CTB required for "error-free" operation of analog systems are about −65 dBc [1], and they can be achieved with κL values from 0.5 to 1.5. A reasonable single-mode yield still exists for such κL values. The application of AR or high reflection (HR) coatings on one or both facets generally implies stronger spatial hole burning. However, an AR coating will also increase the single-mode yield or, equivalently, allow a smaller κL value for the same yield. The AR coating, together with a smaller κL value, has been found to result in small distortion as well [29,32].

We finally note that the optimization of laser structures depends on the specific analog application for which they are intended. It also depends on the used emission wavelength. The leakage currents are, for example, less important in 1.55-μm lasers,

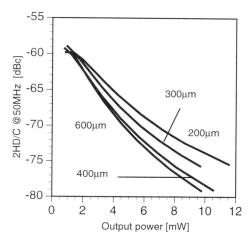

Figure 8.21 Second-order distortion as a function of total bias output power for a cleaved DFB laser with $\kappa L = 1$ and for different cavity lengths. $m = 0.35$, modulation frequency = 50 MHz.

but such lasers usually also need to exhibit a smaller chirp due to the dispersion of standard optical fiber at this wavelength [33]. The second-order distortion is mainly important in CATV systems. The third-order distortion needs to be limited in microwave links between base stations for mobile communications.

References

[1] Darcie, T. E., "Subcarrier Multiplexing for Lightwave Networks and Video Distribution Systems," *IEEE J. Select. Areas Commun.*, Vol. 8, September 1990, pp. 1240–1248.

[2] Cooper, A. J., "Fibre/Radio for the Provision of Cordless/Mobile Telephony Services in the Access Network," *Electron. Lett.*, Vol. 26, November, 1990, pp. 2054–2056.

[3] Morita, K., and H. Ohtsuka, "The New Generation of Wireless Communications Based on Fiber-Radio Technologies," *IEICE Trans. Commun.*, Vol. E76-B, September 1993, pp. 1061–1068.

[4] Okezi, T., and E. Hara, "Measurement of Nonlinear Distortion in Photodiodes," *Electron. Lett.*, Vol. 12, February 1976, pp. 80–81.

[5] Stubkjaer, K., and M. Danielsen, "Nonlinearities of GaAlAs Lasers—Harmonic Distortion," *IEEE J. Quant. Electron.*, Vol. 16, May 1980, pp. 531–537.

[6] Lau, K., and A. Yariv, "Intermodulation Distortion in a Directly Modulated Semiconductor Injection Laser," *Appl. Phys. Lett.*, Vol. 45, November 1984, pp. 1034–1036.

[7] Darcie, T., and R. Tucker, "Intermodulation and Harmonic Distortion in InGaAsP Lasers," *Electron. Lett.*, Vol. 21, August 1985, pp. 665–666.

[8] Morthier, G., "Design and Optimisation of SL-MQW Lasers for High-Speed Analog Communications," *IEEE J. Quant. Electron.*, Vol. 30, July 1994, pp. 1520–1528.

[9] Olshansky, R., P. Hill, V. Lanzisera, and W. Powazinik, "Frequency Response of 1.3 μm InGaAsP High Speed Semiconductor Lasers," *IEEE J. Quant. Electron.*, Vol. 23, September 1987, pp. 1410–1418.

[10] Tucker, R., "High-Speed Modulation of Semiconductor Lasers," *IEEE J. Lightwave Tech.*, Vol. 3, December 1985, pp. 1180–1192.

[11] Morthier, G., F. Libbrecht, K. David, P. Vankwikelberge, and R. Baets, "Theoretical Investigation of the 2nd Order Harmonic Distortion in the AM-Response of 1.55μm F-P and DFB Lasers," *IEEE J. Quant. Electron.*, Vol. 27, June 1991, pp. 1714–1723.

[12] Morthier, G., "Influence of the Carrier Density Dependence of the Absorption on the Harmonic Distortion in Semiconductor Lasers," *IEEE J. Lightwave Tech.*, Vol. 11, January 1993, pp. 16–19.

[13] LaCourse, J., and R. Olshansky, "Observation of Strong Carrier Density Increase Above Threshold and Its Effect on P-I and I-V Characteristics of 1.3μm InGaAsP Lasers," *Proc. IEEE Laser Conference*, Boston, September 1988, pp. 206–207.

[14] Joindot, I., and J.-L. Beylat, "Intervalence Band Absorption Coefficient Measurements in Bulk Layer, Strained and Unstrained Multiquantum Well 1.55μm Semiconductor Lasers," *Electron. Lett.*, Vol. 29, April 1993, pp. 604–606.

[15] Asada, M., A. Kameyama, and Y. Suematsu, "Gain and Intervalence Band Absorption in Quantum-Well Lasers," *IEEE J. Quant. Electron.*, Vol. 20, July 1984, pp. 745–750.

[16] Takemoto, A., H. Watanabe, Y. Nakajima, Y. Sakakibara, S. Kakimoto, and H. Namizaki, "Low Harmonic Distortion Distributed Feedback Laser Diode and Module for CATV Systems," *Proc. Opt. Fiber Communications Conference (OFC'90)*, San Francisco, 1990, p. 214.

[17] Beylat, J.-L., J.-P. Hebert, F. Brillouet, J.-G. Provost, A. Bodere, P. Pagnod-Rossiaux, M. Matabon, Y. Cretin, and M. Hajj, "Very Reproducible 1.55 mm DFB Laser Source With Simultaneous Low Chirp and High Linearity for CATV Distribution Systems," *Proc. ECOC'92*, Berlin, 1992, pp. 895–898.

[18] Haisch, H., J. Bouayad, U. Cebulla, M. Klenk, G. Laube, H. P. Mayer, R. Weinmann, P. Speier, and

E. Zielinski, "Record Performance of 1.55 mm Strained-Layer MQW DFB Lasers for Optical Analog TV Distribution Systems," *Proc. ECOC'92*, Berlin, 1992, pp. 899–902.

[19] Kawamura, H., K. Kamite, H. Yonetani, S. Ogita, H. Soda, and H. Ishikawa, "Effect of Varying Threshold Gain on Second Order Intermodulation Distortion in Distributed Feedback Lasers," *Electron. Lett.*, Vol. 26, September 1990, pp. 1720–1721.

[20] Kinoshita, J., and K. Matsumoto, "Transient Chirping in Distributed Feedback Lasers: Effect of Spatial Hole-Burning Along the Laser Axis," *IEEE J. Quant. Electron.*, Vol. 24, November 1988, pp. 2160–2169.

[21] Buus, J., "Mode Selectivity in DFB Lasers With Cleaved Facets," *Electron. Lett.*, Vol. 21, 1985, pp. 179–180.

[22] Mols, P., P. Kuindersma, M. Van Es-Spiekman, and I. Baele, "Yield and Device Characteristics of DFB Lasers: Statistics and Novel Coating Design in Theory and Experiment," *IEEE J. Quant. Electron.*, Vol. 25, June 1989, pp. 1303–1319.

[23] Lin, M. S., S. Wang, and N. Dutta, "Frequency Dependence of the Harmonic Distortion in InGaAsP Distributed Feedback Lasers," *Proc. Opt. Fiber Communications Conference (OFC'90)*, San Francisco, 1990, p. 213.

[24] Lin, M. S., S. J. Wang, and N. K. Dutta, "Measurements and Modeling of the Harmonic Distortion in InGaAsP Distributed Feedback Lasers," *IEEE J. Quant. Electron.*, Vol. 26, 1990, pp. 998–1004.

[25] Daly, J., "Fiber Optic Intermodulation Distortion," *IEEE Trans. Communications*, Vol. 30, 1982, pp. 1954–1958.

[26] Olshansky, R., "SCM Techniques for Video Distribution, Basic Principles and Options for Upgrades," *Proc. ECOC'90*, Amsterdam, 1990, pp. 855–885.

[27] Saleh, A., "Fundamental Limit on Number of Channels in Subcarrier-Multiplexed Lightwave CATV System," *Electron. Lett.*, Vol. 25, June 1989, pp. 776–777.

[28] Shi, Q., R. S. Burroughs, and D. Lewis, "An Alternative Model for Laser Clipping-Induced Nonlinear Distortion for Analog Lightwave CATV Systems," *IEEE Phot. Tech. Lett.*, Vol. 4, July 1992, pp. 784–787.

[29] Yonetani, H., I. Ushijima, T. Takada, and K. Shima, "Transmission Characteristics of DFB Laser Modules for Analog Applications," *IEEE J. Lightwave Tech.*, Vol. 11, January 1993, pp. 147–153.

[30] Kito, M., H. Sato, N. Otsuka, N. Takenaka, M. Ishino, and Y. Matsui, "The Effect of Dynamic Spatial Hole Burning on the Second and Third Order Intermodulation Distortion in DFB Lasers," *Proc. IEEE Laser Conference'94*, Maui, September 1994, pp. 69–70.

[31] Morthier, G., "The Design of Laser Diodes With Low Distortion for Application in Large Bandwidth CATV Systems," *Summer Topical 95*, Colorado, August 1995.

[32] Okuda, T., H. Yamada, T. Torikai, and Y. Uji, "DFB Laser Intermodulation Distortion Analysis Taking Longitudinal Electrical Field Distribution Into Account," *IEEE Phot. Tech. Lett.*, Vol. 6, January 1994, pp. 27–30.

[33] Philips, M. R., T. E. Darcie, D. Marcuse, G. E. Bodeep, and N. J. Frigo, "Nonlinear Distortion Generated by Dispersive Transmission of Chirped Intensity-Modulated Signals," *IEEE Phot. Tech. Lett.*, Vol. 3, May 1991, pp. 481–483.

Chapter 9

Noise Characteristics of DFB Laser Diodes

This chapter describes the influence of the spontaneous emission and other noise on the power spectrum. This influence is twofold. On one hand, it causes phase fluctuations and a broadening of the emission lines (the width of the emission line corresponding with the main mode is usually called the *laser linewidth*), while on the other hand it also results in fluctuations in the light intensity (expressed by the RIN).

The linewidth defines the coherence of the emitted light for single-mode lasers. As mentioned in Chapter 6, a high degree of coherence is desirable for lasers applied in spectroscopy and most other measurements, in LIDAR systems, and coherent communications. The linewidth limits the resolution in measurement and LIDAR systems and determines the minimum modulation frequency in coherent communications. The RIN, on the other hand, is important in direct optical communication, both digital and analog systems, in which the signal identification at the receiver end is based on the intensity level of the received lightwave. The intensity noise limits the sensitivity of the system (i.e., the minimum received power level that allows signal identification) in such systems.

An important part of this chapter is devoted to possible explanations for the linewidth rebroadening or saturation at high power levels. Indeed, according to well-accepted theories (see [1–3] or Chapter 2), the linewidth ought to decrease with increasing power level, whereas experiments show that this is no longer the case at moderate or high power levels. Today, several explanations for this phenomenon have been brought forward. Which one is valid to explain the linewidth rebroadening in a certain laser rather depends on the specific laser structure. Before dealing with this topic, we will first describe the more general aspects of the noise in DFB lasers, starting with the techniques to measure the noise characteristics. Also, some recent modifications of the rate equation theory of noise will be discussed.

We finally note that an astonishing reduction of the DFB laser linewidths has been witnessed over the past 10 years. We will give an overview of some of the most remarkable results and the methods used to achieve them at the end of the chapter.

9.1 MEASURING NOISE CHARACTERISTICS

The measurement of the intensity noise spectrum can be done with a photodetector (where the intensity fluctuations are converted into electrical current fluctuations) and an electrical spectrum analyzer (to analyze the electrical fluctuations). The intensity noise of laser diodes is usually very low and may give rise to electrical current fluctuations that are below the shot noise level of the detector. A possible solution to this problem is the use of two photodetectors with equal sensitivity, as in the setup of Figure 9.1 [4]. In this setup, the current noise caused by light intensity noise is correlated for both detectors, while the shot noise of both photodetectors is uncorrelated. A data processor can then be used to analyze the correlation of the outputs of the two detectors and identify the light intensity noise in this way.

The measurement of the FM noise spectrum can be based on the detection of the field phase fluctuations using a Michelson interferometer (Figure 9.2). The light intensity $I(t)$ detected by the photodetector in Figure 9.2 is given by

$$I(t) = C[1 + \cos(2\pi f_0 \tau) + \phi(t + \tau) - \phi(t)] \qquad (9.1)$$

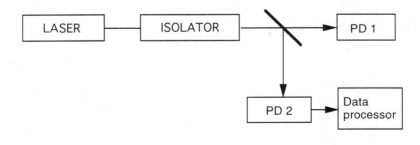

Figure 9.1 Setup for the measurement of the intensity noise, with elimination of the influence of shot noise.

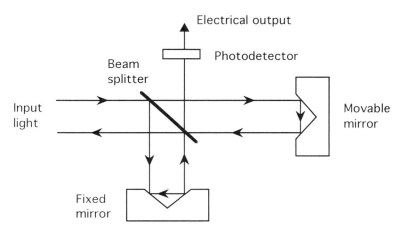

Figure 9.2 Michelson interferometer for the detection of FM noise.

with f_0 the optical center frequency and $\tau = \Delta l/c$ the time delay due to the optical path length difference Δl. If f_0 (or τ) is chosen such that $2f_0\tau = n + 1/2$ (n = integer) and $|\Delta\phi| = |\phi(t + \tau) - \phi(t)| \ll 1$ is assumed, then

$$I(t) = C[1 + \Delta\phi(\tau)] \qquad (9.2)$$

The analysis of the detected signal with an electrical spectrum analyzer allows the determination of the spectrum of the field phase fluctuations $S_{\Delta\phi}(f)$. The FM noise spectrum follows from the phase noise spectrum through the following transformation [4]:

$$S_{\Delta\omega}(f) = \frac{1}{\tau^2}\left[\frac{\pi f \tau}{\sin(\pi f \tau)}\right]^2 S_{\Delta\phi}(f) \qquad (9.3)$$

An alternative is to use a Fabry-Perot interferometer. By adjusting the center frequency f_0 of the laser to one of the points with maximum slope for the transmittance curve of the interferometer, one obtains a conversion of the frequency fluctuations into intensity fluctuations. These can be analyzed using a photodetector and an electrical spectrum analyzer.

For the measurement of the field (power) spectrum, there are two options. One is the straightforward analysis of the optical spectrum using a grating monochromator or an interferometer. As discussed in Chapter 6, this gives typical resolution limits of 10 GHz and 10 MHz for the monochromator and the interferometer, respectively.

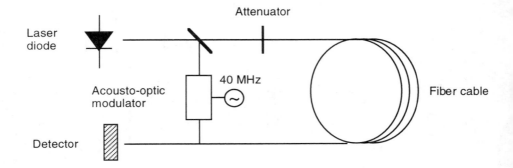

Figure 9.3 Setup for the delayed self-heterodyne method.

A method that allows much better resolution is the delayed self-heterodyne detection method [5] shown in Figure 9.3. The output power of the laser is divided into two parts with different propagation paths. One part of the light is propagating through a long single-mode fiber and is thus delayed by a time τ_d. The other part is frequency-shifted (with an amount f_s) using an acousto-optic modulator. The two signals are mixed again in the photodetector whose output, which contains a signal at the intermediate frequency f_s, is analyzed with an electrical spectrum analyzer.

The detection method relies on the fact that both beams become uncorrelated if the delay time τ_d is much larger than the coherence time of the laser light. The light coming out of the fiber delay line can then be regarded as that from an independent local oscillator, which is as noisy as the signal.

9.2 FM NOISE IN DFB LASERS

9.2.1 Frequency Dependence of the FM Noise Spectrum

An expression for the low-frequency limit of the spectral density of the FM noise was derived in Chapter 2 from the rate equations. This expression can easily be generalized to arbitrary frequencies. From a small-signal solution of the rate equations, one obtains for the fluctuation of the optical pulsation $\Delta\omega$,

$$\Delta\omega = \frac{\alpha}{2}\frac{\partial G}{\partial N}\frac{\left(j\Omega + \frac{\partial G}{\partial S}S_0\right)F_N - \left(j\Omega + G + 2\frac{\partial G}{\partial S}S_0\right)F_S}{\left(j\Omega + \frac{1}{\tau_d} + \frac{\partial G}{\partial N}S_0\right)\left(j\Omega + \frac{\partial G}{\partial S}S_0\right) + \left(G + \frac{\partial G}{\partial S}S_0\right)\left(\frac{\partial G}{\partial N} - \frac{\partial \gamma}{\partial N}\right)} + F_\varphi \quad (9.4)$$

F_S, F_N, and F_φ are now representing the Fourier transforms of the Langevin functions and have the following second-order moments:

$$\langle F_S(\Omega) F_S(\Omega') \rangle = 2\pi \frac{2R_{sp}S}{V_{act}} \delta(\Omega - \Omega')$$

$$\langle F_N(\Omega) F_N(\Omega') \rangle = 2\pi \frac{2}{V_{act}} (AN + BN^2 + CN^3) \delta(\Omega - \Omega')$$

$$\langle F_\varphi(\Omega) F_\varphi(\Omega') \rangle = 2\pi \frac{R_{sp}}{2SV_{act}} \delta(\Omega - \Omega') \qquad (9.5)$$

From now on we will neglect the influence of the carrier shot noise (F_N) and assume that the internal loss γ is independent of the carrier density. The following spectral density of the FM-noise is derived in this case (with $\Omega = 2\pi f$):

$$S_{\Delta\omega}(f) = \frac{R_{sp}}{2S_0 V_{act}} \left\{ 1 + \frac{\alpha^2 \left(\frac{\partial G}{\partial N}\right)^2 \left[\Omega^2 + \left(G + 2\frac{\partial G}{\partial S} S_0\right)^2\right] S_0^2}{(\Omega^2 - \Omega_r^2)^2 + \Omega^2 \theta^2} \right\} \qquad (9.6)$$

with Ω_r and θ the resonance pulsation and damping of the relaxation oscillations:

$$(\Omega_r)^2 = G_{th} \frac{\partial G}{\partial N} S_0$$

$$\theta = \frac{1}{\tau_d} + \frac{\partial G}{\partial N} S_0 + \frac{\partial G}{\partial S} S_0 \qquad (9.7)$$

The expression (9.6) is depicted in Figure 9.4. The presence of relaxation oscillations is clearly visible in the spectrum of the FM noise. It can also be seen that the FM noise spectrum is not white (i.e., flat over the entire frequency range) as has been assumed in Chapter 2 to derive the power spectrum. The low-frequency limit, as formulated in Chapter 2, follows from (9.6) by taking $\Omega = 0$ and neglecting the gain suppression.

9.2.2 Methods for the Calculation of the FM Noise of Complex Laser Structures

The expression (9.6) has been derived from the simple rate equations of Chapter 2 and is therefore not really valid for DFB lasers or even Fabry-Perot lasers with low

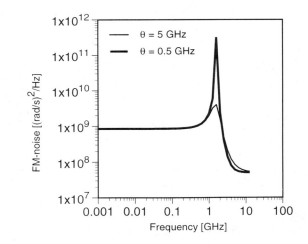

Figure 9.4 Typical FM-noise spectrum of a semiconductor laser.

facet reflectivities. To find the FM noise spectrum in this case, one can turn to the longitudinal rate equations and try to solve them either analytically or numerically. An analytical solution is easily found for Fabry-Perot lasers, but has not yet been reported for DFB lasers. In the case of Fabry-Perot lasers and low frequencies, the problem reduces to finding the variation of the average carrier density. For this purpose, the equations for the number of photons per unit length, s^+ and s^- for the forward and backward propagating waves, can be written as

$$v_g \frac{\partial \ln(s^+)}{\partial z} - 2\Delta\beta_i v_g = \frac{F_s^+(z,t)}{s^+(z,t)}$$

$$-v_g \frac{\partial \ln(s^-)}{\partial z} - 2\Delta\beta_i v_g = \frac{F_s^-(z,t)}{s^-(z,t)} \quad (9.8)$$

After summation of both equations and integration over the laser length, one finds

$$\frac{v_g}{2} \ln\left(\frac{s^+(L)s^-(0)}{s^-(L)s^+(0)}\right) - 2v_g \int_0^L \Delta\beta_i[N(z)] \, dz = \frac{1}{2} \int_0^L dz \left(\frac{F_s^+}{s^+} + \frac{F_s^-}{s^-}\right) \quad (9.9)$$

with $s^-(L) = R_2 s^+(L)$ and $s^+(0) = R_1 s^-(0)$. The fluctuation of the average carrier density is given by

$$\frac{\partial G}{\partial N}\Delta N_0 = -\frac{1}{2L}\int_0^L dz\left(\frac{F_s^+}{s^+} + \frac{F_s^-}{s^-}\right) \qquad (9.10)$$

A more exact expression for the second-order moment of F_φ, which can be derived directly from the coupled-wave equations, is

$$\langle F_\varphi(t)F_\varphi(t')\rangle = \frac{R_{sp}}{8L^2}\int_0^L dz\left(\frac{1}{s^+} + \frac{1}{s^-}\right)\delta(t-t') \qquad (9.11)$$

The rate equation (2.8) then leads to the following low-frequency limit for the spectral density of the FM noise:

$$S_{\Delta\omega}(f) = \frac{R_{sp}}{8L^2}(1+\alpha^2)\int_0^L dz\left(\frac{1}{s^+(z)} + \frac{1}{s^-(z)}\right) \qquad (9.12)$$

An interesting concept that has been introduced to calculate the linewidth of DFB lasers (or more generally any laser that is more complex than a Fabry-Perot laser) is that of the effective linewidth enhancement factor [6]. This concept takes into account the variation of the optical fields along the cavity and leads to expressions for the FM noise of DFB and other complex lasers that are similar to (9.6). At low frequencies, one has

$$S_{\Delta\omega}(f \to 0) = \frac{R_{sp}}{2S_0 V_{act}}(1+\alpha_{eff}^2)K \qquad (9.13)$$

The effective linewidth enhancement factor α_{eff} is given by

$$\alpha_{eff} = \frac{\int_0^L (\alpha\Gamma' + \Gamma'')S(z)\,dz}{\int_0^L (\Gamma' - \alpha\Gamma'')S(z)\,dz} \qquad (9.14)$$

with $S(z)$ the photon density and Γ a longitudinal, complex confinement factor

$$\Gamma(z) = \Gamma'(z) + j\Gamma''(z) = \frac{R^+(z)R^-(z)}{\frac{1}{L}\int_0^L R^+(z)R^-(z)\,dz} \qquad (9.15)$$

and R^+ and R^- the complex field amplitudes of forward and backward propagating waves of the mode under consideration. K is the longitudinal Petermann factor, which also depends on the complex field amplitudes [7,8]

$$K = \left[\frac{\int_0^L |E^2|\, dz}{\left|\int_0^L E^2\, dz\right|} \right]^2 = \left[\frac{\int_0^L (|R^+|^2 + |R^-|^2)\, dz}{\left|\int_0^L R^+R^-\, dz\right|} \right]^2 \qquad (9.16)$$

Expression (9.14) is valid only if spatial hole burning can be neglected, but it is easily extended to cases with spatial hole burning. Figure 9.5 shows the quantity $(1 + \alpha_{\text{eff}}^2)/(1 + \alpha^2)$ for AR-coated DFB lasers as a function of κL.

A second method that has been proposed to calculate the FM noise of DFB lasers or other complex laser structures is based on the solution of the Helmholtz equation using Green's functions [9–12]. The Fourier transform of the electrical field $E(r)$ is related to the source $F(r)$ and the Green's function $G(r, r')$ by

$$E(\mathbf{r}) = \int G(\mathbf{r}, \mathbf{r}')F(\mathbf{r}')\, d\mathbf{r}' \qquad (9.17)$$

If the laser waveguide only sustains the lowest order TE mode ϕ_0, it is possible to write $G(r, r')$ as

$$G(\mathbf{r}, \mathbf{r}') = g(z, z')\phi_0(x)\phi_0(x') \qquad (9.18)$$

with x the lateral, transverse coordinate and with g obeying the equation

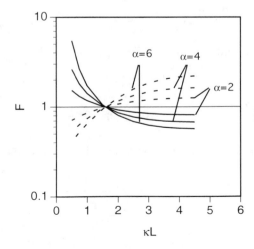

Figure 9.5 $(1 + \alpha_{\text{eff}}^2)/(1 + \alpha^2)$ for AR-coated DFB lasers as a function of κL: (solid line) $\lambda > \lambda_{\text{Bragg}}$; (dashed line) $\lambda < \lambda_{\text{Bragg}}$.

$$\left[\frac{d^2}{dz^2} + k^2\right] g(z, z') = \delta(z - z') \tag{9.19}$$

k is the complex wave number (e.g., including contributions from the carrier density, grating). The solution of the one-dimensional equation (9.19) is given in [13]

$$g(z, z') = \frac{Z_+[\max(z, z')] Z_-[\min(z, z')]}{\dfrac{dZ_+}{dz} Z_- - \dfrac{dZ_-}{dz} Z_+} \tag{9.20}$$

with Z_+ and Z_- fulfilling the homogeneous part of (9.19) and the boundary conditions at positive and negative z, respectively. It can be shown that $Z_+ \sim Z_-$ for a lasing mode, while the normalization of both functions can be chosen such that even $Z_+ = Z_- \equiv Z_0$.

The Green's function method for the calculation of the FM noise consists of the calculation of the field fluctuations using an appropriate Langevin function for the source function F. For a laser with uniform active region, one finds that all cavity effects can be taken into account with the following modification of the spontaneous emission rate R_{sp}

$$R_{sp} = G n_{sp} \left[\frac{\int_0^L |Z_0|^2 \, dz}{\left|\int_0^L Z_0^2 \, dz\right|}\right]^2 \tag{9.21}$$

The last factor of the right-hand side of (9.21) is equal to the longitudinal K-factor.

9.3 LINEWIDTH OF DFB LASERS

In Chapter 2 we derived the laser linewidth from the spectrum of the FM noise under the assumption that this spectrum was white (i.e., constant over the entire frequency range). As we have just seen, this is not really true and a more exact expression for the power spectrum can be based on the expression

$$S_E(f) \sim \int_{-\infty}^{+\infty} dt \, \exp[j 2\pi (f - f_0) t] \, \exp\left[-0.5 \left\langle \left(\int_0^t \Delta\omega(t') \, dt'\right)^2 \right\rangle\right] \tag{9.22}$$

with

$$\left\langle \left(\int_0^t \Delta\omega(t') \, dt'\right)^2 \right\rangle = \frac{1}{2\pi} \int_0^t dt_1 \int_0^t dt_2 \int_{-\infty}^{+\infty} d\Omega \, \langle \Delta\omega(\Omega) \Delta\omega^*(\Omega) \rangle \exp[j \Omega (t_1 - t_2)]$$

This last integral can be calculated from (9.6) with an inverse Fourier transform. One finds

$$\left\langle \left(\int_0^t \Delta\omega(t') \, dt' \right)^2 \right\rangle = \frac{R_{sp}}{2S_0 V_{act}} \left\{ (1 + \alpha^2 A)t + \frac{\alpha^2 [A \cos 3\delta + B \cos \delta]}{\theta \cos \delta} \right. $$
$$\left. - \frac{\alpha^2 e^{-\theta t/2} [A \cos(\Omega_m t - 3\delta) + B \cos(\Omega_m t - \delta)]}{\theta \cos \delta} \right\} \quad (9.23)$$

with

$$\Omega_m^2 = \Omega_r^2 - \frac{\theta^2}{4}, \quad \cos \delta = \frac{\Omega_r}{\sqrt{\Omega_r^2 + \frac{\theta^2}{4}}}$$

$$A = \left(\frac{\partial G}{\partial N} \right)^2 S_0^2 \frac{\left(G + \frac{\partial G}{\partial S} S \right)^2}{\Omega_r^4}, \quad B = \left(\frac{\partial G}{\partial N} \right)^2 S_0^2 \frac{1}{\Omega_r^2} \quad (9.24)$$

The function (9.23) is displayed in Figure 9.6 for $A = 1$, $B = 0$, $\alpha = 4$, $\Omega_r = 10$ GHz, $\theta = 0.5$ GHz, and $R_{sp}/2S_0 V_{act} = 50$ MHz. The oscillations represented by the cosine terms in (9.23) are clearly visible for the case of $\theta = 0.5$ GHz. For more realistic damping constants (e.g., 2 GHz), one can only distinguish weak oscillations.

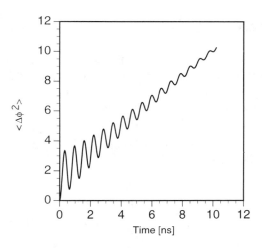

Figure 9.6 Second-order moment of the field phase fluctuations versus time.

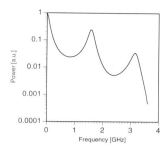

Figure 9.7 Power spectrum resulting from the phase fluctuations of Figure 9.6.

The oscillations contribute little to the 3-dB linewidth. They occur on a frequency scale that is large compared with this linewidth and therefore only affect the tails of the power spectrum. This can be calculated by substituting (9.23) into (9.22), but the integration must be done numerically using an FFT routine. The power spectrum is shown in Figure 9.7 for the $A = 1$, $B = 0$, $\alpha = 4$ $\Omega_r = 10$ GHz, and $\theta = 0.5$ GHz. Side peaks at multiples of the relaxation oscillation frequency can be seen in this spectrum. They were first observed by Daino et al. [14] and Vahala [15].

9.4 CAUSES OF LINEWIDTH REBROADENING IN DFB LASERS

9.4.1 The Presence of Side Modes

The influence of a side mode (with photon density S_1) on the linewidth of the main mode (with photon density S_0) can, in DFB lasers, have its origin in two nonlinearities:

- Spectral hole burning (or, more generally, gain suppression) expressed by the coefficients ξ_{mk} in the rate equations.
- Spatial hole burning, by which, in DFB lasers, mainly the losses γ_0 and γ_1 of main and side modes are affected. These losses depend on the longitudinal variation of the carrier density, which in turn depends on the intensities S_0 and S_1 of both modes.

The interaction via spectral hole burning is easily estimated analytically. In this case, we assume that $\xi_{00} = \xi_{11}$ and $\xi_{01} = \xi_{10} < \xi_{00}$. Indeed, from calculations of the spectral hole burning [16] it seems that the coefficients ξ_{mk} mainly depend on the wavelength difference $|\lambda_m - \lambda_k|$ and that they reach a maximum value for $m = k$. The maximum value then depends only weakly on the wavelength λ_m. From the analytical solution of the rate equations (with two modes being taken into account), it follows that one can distinguish between three regimes [17].

1. The side mode is far below threshold (i.e., strongly suppressed) and one has

$$\frac{\gamma_1}{G_1} - 1 \gg \xi_{00} S_1 \text{ or } \frac{R_{sp}}{V_{act} G_1} \gg \xi_{00} S_1^2 \qquad (9.25)$$

The linewidth of the main mode is independent of the side mode intensity in this case. The fluctuations in the intensity of the side mode are too small to affect the gain via spectral hole burning here. Indeed, the second-order moment of the Langevin function $F_{S,1}$ is proportional to the side-mode intensity and thus very small. Since the side mode is far below threshold, its gain is far less than its loss. The spontaneous emissions, therefore, are hardly amplified and only give small fluctuations in the side-mode intensity. Indeed, from the "static" rate equation for the average photon density of the side mode,

$$(G_1 - \gamma_1)S_1 + \frac{R_{sp}}{V_{act}} + F_{S,1}(t) = 0 \qquad (9.26)$$

It follows that the fluctuations of S_1 are given by

$$\Delta S_1 = \frac{F_{S,1}}{\gamma_1 - G_1} \text{ and } \langle \Delta S_1(t) \Delta S_1(t') \rangle = \frac{2R_{sp} S_1}{V_{act}(\gamma_1 - G_1)^2} = \frac{2S_1^3 V_{act}}{R_{sp}} \qquad (9.27)$$

2. The side mode reaches the threshold, but still has a small intensity. From linearization of the rate equations for S_0 and S_1, one then finds the following approximation for the main-mode linewidth

$$\Delta \nu = \frac{R_{sp}}{4\pi S_0 V_{act}}(1 + \alpha^2) + \frac{\alpha^2 G_1^2 S_1^3 (\xi_{00} - \xi_{01})^2 V_{act}}{4\pi R_{sp}} \qquad (9.28)$$

In this regime, an increase of the linewidth occurs when the side-mode intensity increases as it reaches the threshold. From (9.28), it can be derived that the side mode only has a significant influence when its average intracavity power becomes ±50 µW or more [18,19]. The broadening of the linewidth can be explained by noting that, since the side mode is just below or at threshold, no clamping of the side-mode gain yet exists in this regime. However, the gain of the side mode almost compensates for the loss, and the enhanced resonance results in the strong amplification of all spontaneous emissions. (They are propagated many times back and forth inside the cavity.) Hence, large fluctuations in the side-mode intensity and in the gain (due to spectral hole burning) are generated. Since the gain of the main mode must be clamped to a value equal

to the main mode loss, the fluctuations in the gain, caused by spectral hole burning, must be compensated for by large fluctuations of the carrier density. These large fluctuations in carrier density in turn cause large fluctuations in the refractive index and, as a result of the phase resonance condition, large frequency fluctuations and a large linewidth.
3. The side mode is far above threshold and its intensity no longer consists of ASE. In this case, one finds for the linewidth of the main mode

$$\Delta v = \frac{R_{sp}}{4\pi S_0 V_{act}} \left\{ 1 + \frac{\alpha^2(S_0 + S_1)}{4S_1} \right\} \quad (9.29)$$

The linewidth decreases again with increasing power in main and side modes in this regime. Both modes are now truly oscillating and the fluctuations of the carrier density are restricted by the fact that the gain of both modes must remain equal to the loss. The fluctuations in the intensity of the side mode are more damped as compared with regime 2 and the spectral hole burning effect no longer occurs. The fluctuations of the carrier density are damped by both modes and an increase in the power of one of these modes implies a stronger damping and thus a smaller linewidth.

The influence of the side mode via spatial hole burning can be described in a similar way by expanding the losses γ_0 and γ_1 in the small-signal analysis to

$$\gamma_0 = \gamma_{00}(1 - \sigma_{00}\Delta S_0 - \sigma_{01}\Delta S_1)$$

$$\gamma_1 = \gamma_{00}(1 - \sigma_{10}\Delta S_0 - \sigma_{11}\Delta S_1) \quad (9.30)$$

Numerical values for the coefficients σ_{mk} (which are not necessarily positive numbers) are not easily obtained, however [20]. The effect can be investigated numerically in this case. It is illustrated in Figure 9.8 for a 600-μm-long DFB laser with $\kappa L = 3$ and facet reflectivities $r_1 = 0.566 e^{j\pi}$ and $r_2 = 0.224 e^{j3\pi/2}$. The linewidth has been calculated both with and without taking into account the presence of the side mode. Figure 9.8 also gives the relative side-mode intensity.

Again, one can distinguish between three regimes. As long as the side mode remains strongly suppressed, the fluctuations in its intensity also remain small and the side mode has no influence on the main-mode linewidth. As soon as the side mode approaches the threshold, however, a steep rebroadening of the linewidth occurs. The large fluctuations in the intensity of the side mode now induce large fluctuations in the loss of the main mode via spatial hole burning. Large fluctuations in the carrier density (or the gain), and hence large fluctuations of the refractive index and the

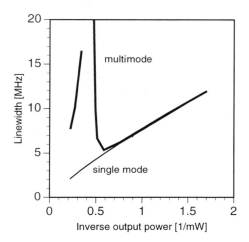

Figure 9.8 Linewidth vs. inverse output power and side-mode suppression for a 600-μm-long DFB laser with $\kappa L = 3$.

frequency, are then needed to compensate for the loss at all times. For the numerical example under consideration, the rebroadening starts when the output power of the side mode reaches ±50 μW, while the output power of the main mode is 2 mW. From other examples, it can be concluded that the rebroadening generally occurs when the side-mode suppression decreases below 20 dB. When the side mode is far above threshold, one again finds a decreasing linewidth for an increasing power level of main and side modes. As for the case of spectral hole burning, the fluctuations of the side-mode intensity are no longer determined by the photon rate equation (which now expresses that the gain and the loss of the side mode must be equal), but by the carrier rate equation, and they remain limited.

9.4.2 Gain Suppression

The influence of gain suppression on the linewidth is mainly a result of the symmetry of the gain suppression around the emission wavelength [21–23]. As a result of this symmetry, there is virtually no influence of gain suppression on the real part of the effective index at the emission wavelength. Indeed, from the Kramers-Krönig relations, it follows immediately that a symmetric gain variation gives rise to an asymmetric refractive index variation. The influence of gain suppression on the differential gain and the absence of any influence on the differential refractive index imply, however, that α must be considered as a bias- or power-dependent quantity. For single-mode lasers, one can write (with $\xi = \xi_{00}$)

$$\alpha = \alpha_0(1 + \xi S_0) \tag{9.31}$$

A gain suppression, as in (2.33), has been used. Substitution of this expression for α in the linewidth formula (2.57) then gives

$$\Delta \nu = \frac{R_{sp}}{4\pi S_0 V_{act}} \{1 + \alpha_0^2(1 + \xi S_0)^2\} \tag{9.32}$$

The bias dependence of α cannot be ignored at high power levels. It gives rise to a minimum in the linewidth at a bias level corresponding with $S_0 = 1/\xi$. The linewidth rebroadening due to the gain suppression is illustrated in Figure 9.9 for a 300-μm-long DFB laser with $\kappa L = 1.5$ and facet reflectivities $r_f = -j0.224$, $r_b = -0.566$. A value of 4.75×10^{-7} [24,25] has been used for ξ. Rebroadening of the linewidth occurs at an output power of about 14 mW.

Experimental evidence for the rebroadening due to spectral hole burning has been given by Park and Buus [26]. They have measured the linewidth at 1-mW output power, the minimum linewidth and the power at which it occurs for two populations of as-cleaved DFB lasers. All lasers were of the buried-ridge type, and the two populations were detuned from the gain peak by 20 and 30 nm, respectively. The two populations therefore had different linewidth enhancement factors, being 7.7 and 6.5,

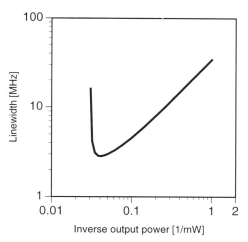

Figure 9.9 Linewidth rebroadening due to gain suppression for a 300-μm-long DFB laser with $\kappa L = 1.5$.

respectively. The experiment showed a good correlation between the linewidth at 1 mW and the linewidth floor. The different gradients for different detuning were consistent with the weak wavelength dependence of the gain suppression factor ξ.

We finally remark that other nonlinearities, which also have a different impact on gain and refractive index, exist. Standing-wave-induced gratings, the simultaneous absorption of two photons generally called two photon absorption, and lateral carrier diffusion can be mentioned here.

9.4.3 Dispersion in the Feedback

The distributed reflections and hence the loss in DFB lasers are strongly wavelength- or frequency-dependent, as has been outlined in Section 2.3 or as can be seen from the wavelength dependence of the round-trip gain. Generally, this dispersion may not be neglected in a linearization of the rate equations. It implies that variations in the frequency (due to modulation or noise) feed back via a variation in the loss. The last variation requires a variation of the carrier density (and thus of the refractive index and the frequency) so that the gain can compensate for the changed loss. Hence, when linearizing the rate equations, one must also include a variation in the loss of the form:

$$\Delta \gamma = \frac{\partial \gamma}{\partial \omega} \Delta \omega + \frac{\partial \gamma}{\partial N} \Delta N \tag{9.33}$$

The variation of the loss with the carrier density is due to, for example, the variation of the average refractive index and of the Bragg wavelength. It accounts for the fact that the loss depends more generally on the Bragg deviation. This variation is also caused by the dispersion (i.e., it is not present if the loss is independent of the Bragg deviation). The equation that expresses the phase resonance must also, due to the dispersion, be transformed into

$$\Delta \omega = \frac{\alpha}{2} \frac{\partial G}{\partial N} \Delta N + \frac{v_g}{2L} \left\{ \frac{\partial \phi_R}{\partial \omega} \Delta \omega + \frac{\partial \phi_R}{\partial N} \Delta N \right\} + F_\varphi(t) \tag{9.34}$$

Expressions for ϕ_R and γ can be derived from the coupled-wave equations. ϕ_R can be interpreted as the phase change associated with distributed and facet reflections. It must be noted that to be exact, one should also include variations of ϕ_R and γ with varying power due to spatial hole burning. Here we concentrate on the dispersion and neglect these variations. The derivatives of γ and ϕ_R can be included in the α-factor. With these adaptations to the rate equations, one can derive the following expression for the linewidth:

$$\Delta\nu = \frac{R_{sp}}{4\pi S_0 V_{act}} \frac{1+\beta^2}{\left\{1 - \frac{v_g}{2L}\frac{\partial\phi_R}{\partial\omega} - \frac{\beta}{2}\frac{\partial\gamma}{\partial\omega}\right\}^2} \quad ; \beta = \frac{\alpha\frac{\partial G}{\partial N} + \frac{v_g}{L}\frac{\partial\phi_R}{\partial N}}{\frac{\partial G}{\partial N} - \frac{\partial\gamma}{\partial N}} \quad (9.35)$$

in which β can be considered as a sort of effective linewidth enhancement factor.

The term between brackets can cause considerable linewidth enhancement when γ and/or ϕ_R are strongly increasing functions of the frequency (i.e., in the case of strong dispersion). Moreover, due to spatial hole burning, this term is also power-dependent. In some lasers where the loss γ and its derivative with respect to frequency increase with bias level, the enhancement increases with bias level. A linewidth floor may then be the result if the increase is sufficiently strong [27,28].

β, $\partial\gamma/\partial\omega$, and $\partial\phi_R/\partial\omega$ are, however, not easily quantified. To illustrate the effect of dispersion, we consider the linewidth of two DFB lasers that are completely identical except for the fact that they have an opposite dispersion. We consider two 300-μm-long, 1.55-μm DFB lasers with field reflection coefficients $r_1 = 0.2324e^{j\pi}$ at the left facet and $r_2 = 0.2324e^{j3\pi/2}$ at the right facet of the first laser (denoted A) and $r_2 = 0.2324e^{j\pi/2}$ at the right facet of the second laser (denoted B). Both lasers have the same longitudinal distribution of optical power (and hence identical spatial hole burning), but also the same threshold gain and threshold gain difference $\Delta gL = 0.33$. Figure 9.10 shows the complex round-trip gain (at threshold) as a function of the wavelength for both lasers. A wavelength-independent gain has been assumed, and hence this wavelength dependence of the round-trip gain is completely due to the dispersion in the loss. The amplitudes of the round-trip gain of both lasers are symmetric with respect to the Bragg wavelength λ_B. The phases, however, are antisymmetric around λ_B.

Both lasers have an opposite dispersion in γ and an equal dispersion in ϕ_R. For example, for laser A (B), it follows from Figure 9.11 that a decrease of λ ($\Delta\omega > 0$) results in an enhanced (reduced) round-trip gain and thus a decreased (increased) loss. Thus $\partial\gamma/\partial\omega$ is negative for laser A and positive for laser B. It can also be seen from Figure 9.10 that since λ_B decreases with increasing N, $\partial\gamma/\partial N$ is positive for laser A and negative for B. Any difference in linewidth between both lasers can be attributed to the dispersion in γ and to the different β-values.

The linewidth as a function of inverse output power is shown in Figure 9.11 for both lasers. The α-factor has a value of 3. The dispersion still causes the linewidth of laser B to be about 15% larger than that of laser A at low power levels, while its influence increases with increasing power level. The linewidth of laser B reaches a minimum of 77 MHz at a power level of about 2 mW and then increases again to a value

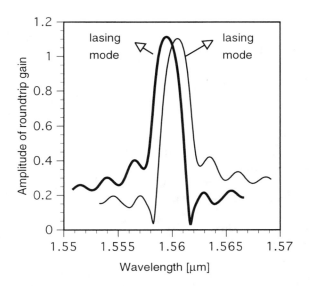

Figure 9.10 Complex round-trip gain at threshold of lasers A (—) and B (━).

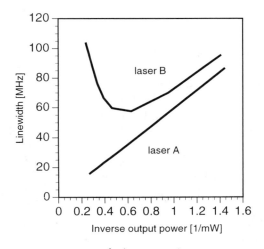

Figure 9.11 Linewidth vs. inverse output power for lasers A and B.

of 120 MHz at a power level of 3.7 mW. Beyond this power level, the laser becomes multimode and a steep increase of the linewidth starts.

This type of rebroadening has theoretically been observed in most lasers where there is a local increase of the loss with the frequency in the neighborhood of the emission frequency and where the loss increases with bias level. The minimum linewidth and the power at which this minimum occurs depend on the strength of the dispersion and of the spatial hole burning, but also on the value of the linewidth enhancement factor. The dispersion is often much weaker than it is in laser B, but the linewidth enhancement factor is often larger (values of 5 and more have been reported).

Since many lasers emitting on the short-wavelength side of the Bragg wavelength seem to become multimode at low power levels (they are unstable), this rebroadening is only found for a limited number of lasers.

The implications of dispersion in the gain can be estimated by replacing $\partial\gamma/\partial\omega$ in (9.34) with $\partial\gamma/\partial\omega - \partial G/\partial\omega$. The dispersion in the gain is generally negligible compared with the dispersion in the loss. The dispersion in the gain affects the linewidth by less than 1%.

The previous results also indicate that dispersion of the loss, even if it does not result in a linewidth rebroadening, can affect the linewidth considerably and should not be ignored.

9.5 RELATIVE INTENSITY NOISE OF DFB LASERS

The intensity noise (i.e., fluctuations in the emitted optical power) is of special importance for the application of analog or digital AM communication. Future applications of this can be found in, for example, optical CATV systems [29]. The CNR has a large influence on the system performance in this case, and it is determined by the noise of the receiver (e.g., shot noise of the photodetector) as well as by the intensity noise of the laser [30].

One must distinguish between the noise, which appears in the intensity of a single mode (e.g., the main mode), and the noise, which appears in the total intensity (equal to the sum of the intensities of all modes). Mode partition noise [31,32], for example, involves large fluctuations in the intensity of each mode separately, while the fluctuations in the total intensity remain rather small. The fluctuations in the intensities of the different modes are thereby correlated so that their sum remains very small. Mode partition noise should nonetheless be avoided in optical communication systems. Indeed, the correlation between the noise in the modes can be destroyed as a result of the dispersion in the optical fiber (which causes all modes to propagate with different velocity), and it can result in large fluctuations of the total intensity measured by the photodetector. This noise is easily reduced by working at the wavelength

(1.3 μm) where minimum fiber dispersion occurs. Noise in the total intensity of the laser itself is not as easily reduced. Theoretical and experimental investigation of the factors that determine the RIN is therefore required.

9.5.1 Frequency Dependence of the RIN in Single-Mode Lasers

Figure 9.12 shows a typical spectrum of the RIN of a single-mode laser. This spectrum is flat for frequencies ranging from 1 to a few hundred megahertz (or more), where it strongly increases as a result of the relaxation oscillation. The spectral density of the RIN again decreases beyond this oscillation.

It can no longer be argued, as for the FM noise, that only the low-frequency value of the RIN has implications for the performance of a communication system. The relaxation oscillations in the FM noise only limit the channel spacing in communication systems. The fiber bandwidth is nevertheless large enough so that a sufficient number of channels are allowed. The relaxation oscillations in the RIN, on the other hand, limit the maximum modulation frequency. A low value of the RIN at these relaxation oscillations would imply that a larger channel bandwidth can be used.

The frequency dependence of the RIN can be derived from the rate equations. For a Fabry-Perot laser, one finds for the intensity fluctuations at high power levels,

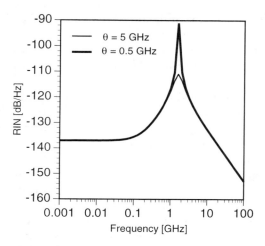

Figure 9.12 Typical RIN spectrum of a single-mode laser.

$$\frac{\Delta S}{S_0} = \frac{\dfrac{\partial G}{\partial N}F_N + \left(j\Omega + \dfrac{1}{\tau_d}\right)\dfrac{F_S}{S_0}}{(\Omega_r^2 - \Omega^2) + j\Omega\theta}$$

$$\mathrm{RIN}\!\left(f = \frac{\Omega}{2\pi}\right) = \frac{2}{V_{act}} \; \frac{\left(\dfrac{\partial G}{\partial N}\right)^{2}(AN_0 + BN_0^2 + CN_0^3) + \left(\Omega^2 + \dfrac{1}{\tau_d^2}\right)\dfrac{R_{sp}}{S_0}}{(\Omega^2 - \Omega_r^2)^2 + \Omega^2\theta^2} \qquad (9.36)$$

From this expression it follows that the low-frequency limit of the RIN decreases with output power (or S_0) as the inverse of the third order of the power at low power levels and as the inverse of the power level at higher power levels.

The damping of the relaxation oscillations is mainly caused by the gain suppression. From (9.36) it can be derived that the maximum value of the RIN, occurring at the resonance frequency, is proportional to the inverse of the square of ξ and to the inverse of the third power of the photon number. An increase of the output power will therefore bring about a considerable decrease of the RIN.

9.5.2 Factors Determining Low-Frequency RIN

Multimode Lasers

The RIN appears to be very sensitive to the presence of strong side modes. This is first of all the case for the noise in the intensity of the main and side modes separately (partition noise), and is easily explained with the help of the rate equations. Equation (2.10) for the side-mode intensity (index 1) yields in the low-frequency approximation,

$$\Delta S_1 = \frac{F_{S,1}(t) + S_1 \Delta[G(N_0, \lambda_1) - \gamma_1]}{\gamma_1 - G(N_0, \lambda_1)} \qquad (9.37)$$

Both the side-mode intensity S_1 and the fluctuations in gain and loss $\Delta[G(N_0, \lambda_1) - \gamma_1]$ are relatively small for a side mode below threshold. The fluctuations in the carrier density are indeed restricted by the oscillation of the main mode and the resulting gain clamping, while nonlinearities such as spectral and spatial hole burning can be neglected in first-order approximation. The second term in the numerator of (9.37) can thus be neglected. The denominator of (9.37), however, is very small when the side mode reaches the threshold. Its value can be determined from the solution of (2.10) in the steady state, and one then finds for the spectrum of the low-frequency intensity noise of the side mode,

$$S_{\Delta S_1}(f) = \frac{2V_{act}S_1^3}{R_{sp}[1 + (2\pi f\tau_1)^2]}, \quad \tau_1 = \frac{V_{act}S_1}{R_{sp}} \quad (9.38)$$

The fluctuations in the side-mode intensity are proportional to the third power of the side-mode intensity, and they can be relatively large if the side mode approaches the threshold.

By neglecting all other Langevin functions and all nonlinearities, it follows from the carrier rate equation that the total intensity remains constant. This indicates that the fluctuations in the side-mode intensity and in the main-mode intensity are cross correlated so that large fluctuations in the main-mode intensity will be present. It is as if the photons are just repartitioned over the two modes in a stochastic way, a viewpoint giving rise to the name *partition noise*.

The other Langevin functions and the fluctuations in gain and carrier density are not zero in reality, but their effect is much smaller than the fluctuations in the side-mode intensity. The mode partition noise is illustrated in Figure 9.13, which gives the fluctuations in the main mode intensity for the 600-μm-long DFB laser with $\kappa L = 3$ that was also used to obtain Figure 9.8. This figure also shows that the influence of the partition noise is restricted to low frequencies (up to 100 MHz). The cutoff frequency is given by $[G(N_0, \lambda_1) - \gamma_1]/2\pi$, as can be seen after solution of (2.10) in the dynamic regime.

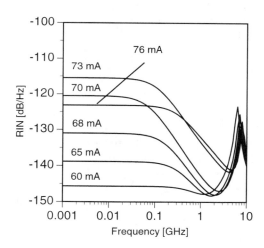

Figure 9.13 RIN of the main mode under the presence of a relatively strong side mode.

The fluctuations in the intensity of main and side modes become again smaller beyond the threshold of the side mode. The side mode then no longer consists of ASE, but becomes a genuine oscillator mode, and the side-mode equation now reduces to the condition that the gain must compensate for the loss of the side mode. An analytical approximation can be found only if spatial or spectral hole burning are considered.

These nonlinearities furthermore result in a strong dependence of the total intensity noise on the side-mode strength [33], in contrast with what has been believed for a long time. As has already been outlined in Section 9.4, the fluctuations in the side-mode intensity induce large fluctuations in the gain or loss of the main mode, which ought to be compensated for by large fluctuations in the carrier density to maintain the oscillation condition. The conservation of charge (expressed by the carrier rate equation) implies that the large fluctuations in carrier density now also require large fluctuations in the total intensity.

Single-Mode Lasers

As long as a strong side-mode suppression (more than 30 dB) exists, the RIN decreases with increasing power, as can be seen from (9.36). Gain suppression and spatial hole burning only have a very weak influence in this case. The dispersion results in a dependence of the RIN on the FM noise. If gain suppression is neglected, one finds

$$G\Delta S(t) = F_N(t) + \frac{\left(\frac{1}{\tau_d} + \frac{\partial \gamma}{\partial N} S_0\right)}{\left(\frac{\partial G}{\partial N} - \frac{\partial \gamma}{\partial N}\right)} \frac{F_S}{S_0} - \frac{\left(\frac{1}{\tau_d} + \frac{\partial G}{\partial N} S_0\right)}{\left(\frac{\partial G}{\partial N} - \frac{\partial \gamma}{\partial N}\right)} \frac{\partial \gamma}{\partial \omega} \Delta\omega(t) \quad (9.39)$$

The RIN decreases as S^{-3} at low power levels. At higher power levels, the shot noise becomes more important; the RIN still decreases, but not as rapidly anymore. It must be noted, however, that the observed RIN at these high power levels is usually dominated by the noise of the photodetector, noise that has not been taken into account here.

The previous results indicate that for practical purposes, a sufficiently low RIN (e.g., −150 dB/Hz) can generally be achieved by biasing at a sufficiently high power level, provided that a sufficient suppression of the external reflections and of the side modes can be guaranteed.

9.6 DESIGNING HIGHLY COHERENT DFB LASERS

From (2.56) or (9.13) for the linewidth, one can already identify the main factors that have to be optimized to design highly coherent DFB lasers. The spontaneous emission rate and hence the threshold gain Γg needs to be as small as possible and the photon number $S_0 V_{act}$ must be as large as possible. If the output power is fixed, then the photon number can be increased by increasing the facet reflectivities, the κL value, and the active layer dimensions. An increase of the facet reflectivities and κL decreases the threshold gain as well, but gives a reduced external efficiency. A compromise is recommended in this case, but the reflectivity of the rear facet (of which the output is not used, unless for monitoring) should still be as high as possible. Increasing the active layer dimensions will also decrease the threshold gain. This is particularly true for the laser length.

The (effective) linewidth enhancement factor can be reduced by so-called detuning. As a function of the wavelength, the linewidth enhancement factor of the active layer material typically increases as shown in Figure 9.14. Therefore, one should force the emission wavelength to be as short as possible. This can be accomplished by choosing a very short grating period in DFB lasers. By such detuning, however, one forces the emission wavelength farther away from the peak wavelength of the gain curve. An increased threshold current and even emission in Fabry-Perot modes (with negligible distributed feedback) can be the consequence, so that again a compromise is recommended.

If quantum-well material is used for the active layer, there is also a dependence of the linewidth enhancement factor on the carrier density. This dependence has its

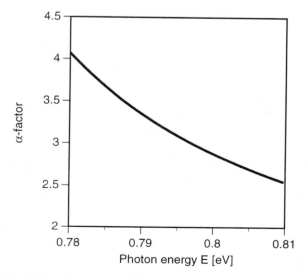

Figure 9.14 Linewidth enhancement factor versus wavelength.

origin in the sublinear dependence of gain on carrier density, while the refractive index remains nearly linear as a function of carrier density. From the relations

$$g = g_0 \ln\left(\frac{N}{N_0}\right)$$

$$n = n_0 + n_1 N \tag{9.40}$$

the following expression for α can be derived

$$\alpha = -\frac{2kn_1}{g_0} N = \alpha_0 \exp\left(\frac{G_{th}}{\Gamma g_0 v_g}\right), \quad \alpha_0 = -\frac{2kn_1}{g_0} N_0 \tag{9.41}$$

It is obvious from this expression that α decreases with decreasing threshold gain and thus with increasing laser length, increasing facet reflectivities, and increasing κL value. An increase of the transverse and lateral dimensions will increase the confinement factor Γ and thus cause a strong decrease of α as well.

The effective linewidth enhancement factor can be reduced by choosing special longitudinal structures. The decrease obtained in this way is, however, relatively small for lasers with cleaved or HR-coated facets.

Most of the very good linewidth results that have been achieved in recent years have been obtained with relatively long lasers and at high power levels (i.e., for relatively high photon numbers). Ogita et al. [34] obtained a linewidth of 830 kHz at 25 mW of output power using a 1,200-μm-long cavity with $\kappa L = 2.2$ and a bulk active layer. Kitamura et al. [35] reported a linewidth of 250 kHz at an output power of 4 mW with a 1,500-μm-long multi-quantum-well DFB laser. Besides the large length, a detuning of −24 nm was also applied so that a very small α-factor resulted. Bissessur et al. [36] obtained a linewidth of 70 kHz at 10 mW using a cavity length of 1,450 μm and a κL of 2.9. The active layer consisted of strained-layer multi-quantum-well material. Very low internal losses (due to the very weak intervalence band absorption) and an α-factor of 1.7 were measured for their lasers.

References

[1] Henry, C., "Theory of the Phase Noise and Power Spectrum of a Single Mode Injection Laser," *IEEE J. Quant. Electron.*, Vol. 19, September 1983, pp. 1391–1397.
[2] Agrawal, G., and N. Dutta, *Long-Wavelength Semiconductor Lasers*, New York: Van Nostrand Reinhold, 1986.
[3] Vahala, K., and A. Yariv, "Semiclassical Theory of Noise in Semiconductor Lasers—Part II," *IEEE J. Quant. Electron.*, Vol. 19, June 1983, pp. 1101–1109.
[4] Okoshi, T., and K. Kikuchi, *Coherent Optical Fiber Communications*, Tokyo: KTK Scientific Publishers/Kluwer Academic Publishers, 1988.

[5] Okoshi, T., K. Kikuchi, and A. Nakayama, "Novel Method for High Resolution Measurement of Laser Output Spectrum," *Electron. Lett.*, Vol. 16, June 1980, pp. 630–631.
[6] Amann, M., "Linewidth Enhancement in Distributed Feedback Semiconductor Lasers," *Electron. Lett.*, Vol. 26, April 1990, pp. 569–571.
[7] Wang, J., N. Schunk, and K. Petermann, "Linewidth Enhancement for DFB Lasers Due to Longitudinal Field Dependence in the Laser Cavity," *Electron. Lett.*, Vol. 23, July 1987, pp. 715–716.
[8] Arnaud, J., "Natural Linewidth of Semiconductor Lasers," *Electron. Lett.*, Vol. 22, May 1986, pp. 538–540.
[9] Henry, C., "Theory of Spontaneous Emission Noise in Open Resonators and Its Application to Lasers and Optical Amplifiers," *IEEE J. Lightwave Tech.*, Vol. 4, March 1986, pp. 288–297.
[10] Kojima, K., and K. Kyuma, "Analysis of the Linewidth of Distributed Feedback Laser Diodes Using the Green's Function Method," *Jap. J. Appl. Phys.*, Vol. 27, September 1988, pp. 1721–1723.
[11] Makino, T., "Analysis of the Spontaneous Emission Rate of Multiple-Phase-Shift Distributed Feedback Semiconductor Lasers," *Electron. Lett.*, Vol. 26, May 1990, pp. 629–630.
[12] Kikuchi, K., and H. Tomofuji, "Analysis of Linewidth of Separated-Electrode DFB Laser Diode," *Electron. Lett.*, Vol. 25, July 1989, pp. 916–918.
[13] Morse, P. M., and H. Feshbach, *Methods of Theoretical Physics*, New York: McGraw-Hill, 1953.
[14] Daino, B., P. Spano, M. Tamburrini, and S. Piazzolla, "Phase Noise and Spectral Line Shape in Semiconductor Lasers," *IEEE J. Quant. Electron.*, Vol. 19, March 1983, pp. 266–270.
[15] Vahala, K., "Corrections to the Rate Equation Approximation for Dynamic Considerations in a Semiconductor Laser," *Appl. Phys. Lett.*, Vol. 48, May 1986, pp. 1340–1341.
[16] Asada, M., and Y. Suematsu, "Density-Matrix Theory of Semiconductor Lasers with Relaxation Broadening Model—Gain and Gain Suppression in Semiconductor Lasers," *IEEE J. Quant. Electron.*, Vol. 21, May 1985, pp. 434–442.
[17] Krüger, U., and K. Petermann, "The Semiconductor Laser Linewidth Due to the Presence of Side Modes," *IEEE J. Quant. Electron.*, Vol. 24, December 1988, pp. 2355–2358.
[18] Miller, S., "The Effect of Side Modes on Linewidth and Intensity Fluctuations in Semiconductor Lasers," *IEEE J. Quant. Electron.*, Vol. 24, May 1988, pp. 750–757.
[19] Adams, M., "Linewidth of a Single Mode in a Multimode Injection Laser," *Electron. Lett.*, Vol. 19, August 1983, pp. 652–653.
[20] Pan, X., B. Tromborg, and H. Olesen, "Linewidth Rebroadening in DFB Lasers Due to Weak Side Modes," *IEEE Phot. Tech. Lett.*, Vol. 3, 1991, pp. 112–114.
[21] Morthier, G., P. Vankwikelberge, F. Buytaert, and R. Baets, "Influence of Gain Non-Linearities on the Linewidth Enhancement Factor in Semiconductor Lasers," *IEE Proc.*, Pt. J, Vol. 137, February 1990, pp. 30–32.
[22] Agrawal, G., "Intensity Dependence of the Linewidth Enhancement Factor and Its Implications for Semiconductor Lasers," *IEEE Phot. Tech. Lett.*, Vol. 1, August 1989, pp. 212–214.
[23] Pan, X., H. Olesen, and B. Tromborg, "Influence of Nonlinear Gain on DFB Laser Linewidth," *Electron. Lett.*, Vol. 26, July 1990, pp. 1074–1076.
[24] Tucker, R., "High-Speed Modulation of Semiconductor Lasers," *IEEE J. Lightwave Tech.*, Vol. 3, December 1985, pp. 1180–1192.
[25] Wiesenfeld, J., R. Tucker, and P. Downey, "Picosecond Measurement of Chirp in Gain-Switched Single-Mode Injection Lasers," *Appl. Phys. Lett.*, Vol. 51, 1986, pp. 1307–1309.
[26] Park, C., and J. Buus, "Prediction of the Linewidth Floor in DFB Lasers," *Proc. Opt. Fiber Communication Conference (OFC'90)*, San Francisco, p. 158.
[27] Morthier, G., K. David, P. Vankwikelberge, and R. Baets, "Linewidth Rebroadening in DFB-Lasers Due to a Bias Dependent Dispersion of the Feedback," *Electron. Lett.*, Vol. 27, February 1991, pp. 375–377.
[28] Olesen, H., B. Tromborg, H. E. Lassen, and X. Pan, "Mode Instability and Linewidth Rebroadening in DFB Lasers," *Electron. Lett.*, Vol. 28, February 1992, pp. 444–446.

[29] Sato, K., "Intensity Noise of Semiconductor Laser Diodes in Fiber-Optic Analog Video Transmission," *IEEE J. Quant. Electron.*, Vol. 19, September 1983, pp. 1380–1391.
[30] Way, W., "Subcarrier Multiplexed Lightwave System Design Considerations for Subscriber Loop Applications," *IEEE J. Lightwave Tech.*, Vol. 7, November 1989, pp. 1806–1818.
[31] Henry, C., P. Henry, and M. Lax, "Partition Fluctuations in Nearly Single-Longitudinal-Mode Lasers," *IEEE J. Lightwave Tech.*, Vol. 2, June 1984, pp. 209–216.
[32] Liou, K., M. Ohtsu, C. Burrus, U. Koren, and T. Koch, "Power Partition Fluctuations in Two-Mode-Degenerate Distributed-Feedback Lasers," *IEEE J. Lightwave Tech.*, Vol. 7, April 1989, pp. 632–639.
[33] Su, C., J. Schlafer, and R. Lauer, "Low Frequency Relative Intensity Noise in Semiconductor Lasers," *Proc. Opt. Fiber Communication Conference (OFC'90)*, San Francisco, 1990, p. 218.
[34] Ogita, S., Y. Kotaki, M. Matsuda, Y. Kuwahara, and H. Ishikawa, "Long-Cavity Multiple-Phase-Shift Distributed Feedback Laser Diode for Linewidth Narrowing," *IEEE J. Lightwave Tech.*, Vol. 8, October 1990, pp. 1596–1603.
[35] Kitamura, M., H. Yamazaki, T. Sasaki, N. Kida, H. Hasumi, and I. Mito, "250 kHz Spectral Linewidth Operation of 1.5μm Multiple Quantum Well DFB-LD's," *IEEE Phot. Tech. Lett.*, Vol. 2, May 1990, pp. 310–311.
[36] Bissessur, H., C. Starck, J.-Y. Emery, F. Pommereau, C. Duchemin, J.-G. Provost, J.-L. Beylat, and B. Fernier, "Very Narrow-Linewidth (70 kHz) 1.55μm Strained MQW DFB Lasers," *Electron. Lett.*, Vol. 28, May 1992, pp. 998–999.

Chapter 10

Fabrication and Packaging of DFB Laser Diodes

A good laser diode design leads nowhere without careful fabrication and packaging of the real component. However, for that reason, a good laser diode design is also a design in which possible difficulties and limitations of the fabrication have been anticipated. We therefore believe that a basic knowledge of the fabrication and packaging of DFB laser diodes is essential for everyone dealing with DFB laser diodes in one way or another.

In this chapter, we will only give a very brief overview of the most important techniques used in the fabrication of DFB laser diodes and of the most relevant packaging issues. It is by no means our intention to give a detailed treatment of the different technological and chemical aspects, let alone to provide the reader with recipes for the successful growth of specific designs. We refer to other, more appropriate textbooks (e.g., see [1–3]) for this.

Essential in the fabrication of semiconductor lasers is the epitaxial or crystalline growth of two semiconductors with nearly equal lattice constant one on top of the other. There are essentially three different techniques to achieve this: liquid-phase epitaxy (LPE), vapor-phase epitaxy (VPE), and molecular beam epitaxy (MBE). These techniques will be described in the first paragraph of this chapter. Essential in the fabrication of DFB lasers, furthermore, is the introduction of a periodic diffraction grating with a small period in the device. The techniques for grating fabrication are described in the second paragraph.

The packaging serves to connect devices with the outside world and protect them from environmental influences (such as humidity, temperature, pressure, and vibrations). In other words, it makes devices more robust and easier to handle and is therefore also of great importance. The connections with the outside world consist of an optical connection (i.e., the coupling of the emitted light into a fiber) and electrical connections through which currents are reaching the electrodes on the chip. The optical coupling requires attention in order to minimize the negative influence of external reflections and to achieve maximum power in the optical fiber. The electrical coupling, on the other hand, mainly has an impact on the dynamic characteristics and must be carefully designed such that the dynamic characteristics of a laser chip are not degraded by inefficient current injection at high frequencies. The packaging is discussed in the third paragraph.

10.1 LASER DIODE FABRICATION TECHNIQUES

The different fabrication steps necessary in the realization of a laser structure are illustrated in Figure 10.1 for a simple structure such as the EMBH [4]. First,

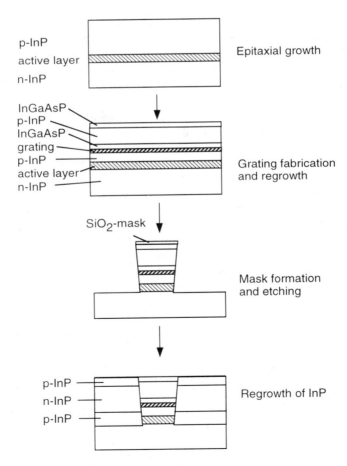

Figure 10.1 Process steps in the fabrication of a DFB laser.

crystalline layers (the InGaAsP active layer and a p-InP layer) are grown on the n-InP substrate using one of the possible hetero-epitaxy techniques. A grating is etched in the p-InP layer and additional epitaxial layers (InGaAsP, p-InP, and an InGaAsP contact layer) are grown on top of this grating. After this, a SiO_2-mask is formed using chemical vapor deposition and photolithography. Different etching processes [5–7] can then be used to etch all the material not covered by the mask away up to the substrate in order to form a mesa. Another epitaxial regrowth (of current blocking layers consisting of, for example, p-InP, n-InP, and p-InP) then finally results in the EMBH structure, on which contacts have to be deposited using, for example, sputtering, plating, or evaporation.

The epitaxial growth of crystalline layers is discussed hereafter. The fabrication of gratings is described in the next section. The deposition of contacts and the different etching processes are not discussed in this book. We refer to more specialized books for more details on these issues.

10.1.1 Liquid-Phase Epitaxy

In this technique, an oversaturated solution of the material to be grown is brought in contact with a crystalline semiconductor layer or substrate for some time. Sedimentation of the material to be grown then results if the substrate and the desired sediment have approximately the same lattice constant. A disadvantage of LPE is that the thinnest layer that can be grown is only about 0.1 µm.

LPE apparatus generally used in the fabrication of laser diodes is of the so-called multibin-boat type, shown in Figure 10.2. It basically consists of a sliding rod or boat in graphite that has a number of reservoirs. The different reservoirs are filled with solutions of the different materials that have to be grown. The substrate is fixed in a groove of a graphite holder. Moving the rod therefore results in the subsequent

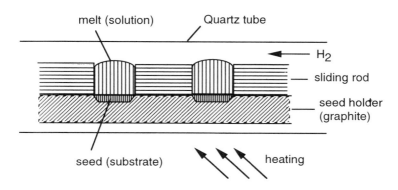

Figure 10.2 Multibin-boat apparatus for LPE growth.

sedimentation from the solutions that are in the subsequent reservoirs. The thickness of the different layers obviously depends on the speed with which the rod is moved. Hydrogen is usually used as an ambient gas to prevent oxidation.

10.1.2 Molecular Beam Epitaxy

The principle of MBE is illustrated in Figure 10.3. Atomic and molecular beams are directed toward a heated substrate under ultrahigh vacuum (UHV) conditions. Their constituents bond with the substrate and give rise to a lattice-matched layer. Other beams besides the In and P_2 beams shown in Figure 10.3 are generally used, such as Ga and P_2 beams, Al-beams, and dopant-beams. The technique allows a precise control of growth rate, composition, and doping density, but gives a rather slow growth.

Conventional solid-source MBE is not very well suited to the growth of long-wavelength optoelectronic devices because of the high vapor pressures of solid arsenic and phosphorus. To obtain a controllable flux of P_2 and As_2, the use of gaseous sources such as phosphine and arsine (obtained through a high-temperature cracker) was proposed by Panish [8] and Calawa [9]. This technique is now commonly known as gas source MBE (GSMBE) and is used more widely in the fabrication of laser diodes.

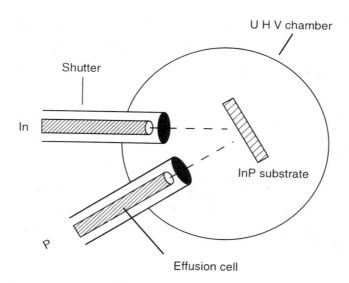

Figure 10.3 Principle of molecular beam epitaxy

10.1.3 Metal-Organic Vapor-Phase Epitaxy

In metal-organic vapor-phase epitaxy (MOVPE), which is the most widely used technique in the fabrication of DFB laser diodes, epitaxial layers are grown from gaseous source chemicals. The principle is illustrated in Figure 10.4. The apparatus consists of an open tube quartz reactor in which metal-organics and hydrides are entered on one side and from which rest gasses leave at the other side. Inside the reactor, a substrate is placed on a heated graphite susceptor. Above this heated susceptor, a reaction between the metal-organic gases (of group III elements or dopants) and the hydrides (of group V elements or dopants) takes place and results in the deposition of a thin monocrystalline layer on the substrate. A carrier gas, very pure hydrogen, is added to the gas flow to transport the source gases into the reactor. The heating is achieved by radiofrequency (RF) induction, infrared radiation, or resistance heating. The temperature in the reactor is typically 600° C. The concentration of the different gases in the reactor is determined by the mass flow of the gases and is controlled using electronic mass flow conrollers.

MOVPE, which is also known as metal-organic chemical vapor deposition (MOCVD), requires a relatively simple growth apparatus and is very versatile. Like GSMBE, however, it makes use of very toxic gases like arsine and phosphine and hence requires precautions to protect operator and environment. Therefore, there has been significant research on substitutes for arsine and phosphine. The most valid

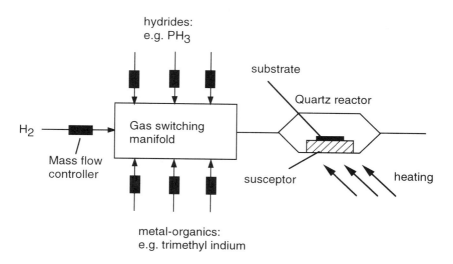

Figure 10.4 Principle of metal-organic vapor-phase epitaxy.

alternatives are liquids with relatively low vapor pressure like TBA (tertiary butyl arsine) and TBP (tertiary butyl phosphine). Due to their high cost, these sources are not yet widely accepted, however.

10.2 Grating Fabrication Techniques

The different growth and processing steps are very similar for Fabry-Perot laser diodes and DFB laser diodes. The formation of a periodic diffraction grating in DFB lasers is obviously an exception to this rule. The most common technique for the fabrication of such a grating is based on holographic exposure of a thin (typically 50 to 100 nm thick) layer of photoresist that is spun onto a wafer surface. The photoresist layer is exposed to two collimated, expanded beams from a blue or ultraviolet laser at an appropriate angle to form high-contrast fringes with the desired period (Figure 10.5). After development, the resist pattern can act as an etch mask for the underlying layers. Etching of the surface as shown in Figure 10.6 finally leads to a grating in InP. Since the wavelength of the used lasers is known precisely and angles in the milliradian range can be measured, the corrugation period (typically around 200 nm for a wavelength of 1.55 µm) can be defined to angstrom-level precision.

Another technique is based on the growth of a thin InGaAsP layer on the InP layer. After the etching of the corrugation and InP regrowth, a rectangular grating consisting of InGaAsP islands surrounded by InP results. This technique allows a better reproducibility, even of gratings with very weak coupling, because it is less sensitive to etch depth and because the InGaAsP layer can be made very thin.

Holographic definition of the diffraction always results in the grating having a sinusoidal shape. Thus, it cannot be used when phase shifts and/or variations in

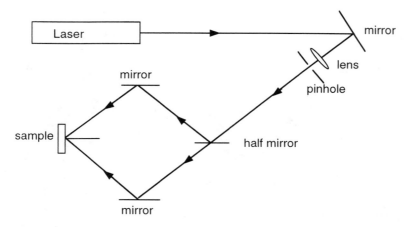

Figure 10.5 Holographic exposure technique for grating fabrication.

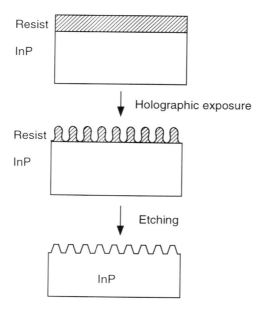

Figure 10.6 Grating fabrication in the photoresist and the InP.

period or amplitude have to be introduced in the grating. An alternative technique, which allows the definition of any grating shape, is electron beam lithography [10]. In this technique, an electron beam writes the desired pattern on an electron beam resist, which after exposure acts as an etch mask. This technique gives much more flexibility than the holographic technique, but has the disadvantage of having longer writing times and/or a limited exposure area. This last restriction requires stitching together a number of exposure areas to pattern larger areas, which may cause small, uncontrollable phase shifts in the grating.

10.3 Packaging of DFB Laser Diodes

As mentioned in the introduction, two main aspects of the packaging are the optical and electrical connections with the outside world. A brief discussion of these aspects will be given in the following sections. There are, however, several other aspects. Components such as a monitoring photodiode, a temperature sensor and a thermoelectric cooler need to be included. The protection against environmental influences such as temperature, humidity, pressure, and shock requires firm positioning of the different optical and optoelectronic components and choosing the right materials. Because of the small alignment tolerances and the heat dissipation of laser diodes, materials with a low thermal expansion coefficient are preferred. The reader should refer to [11–13] for a more extensive account of the packaging.

10.3.1 Electrical Aspects

The electrical path connecting the device with cables is critical for the high-speed operation of the laser diode. Indeed, the high-frequency signal will be reflected if the chip and its connection to the package are not matched to the coaxial cable that connects package and signal generator. The chip and its connection to the package can be regarded as a load Z_L of the coaxial cable. The voltage reflection r on a transmission line with characteristic impedance Z_0 is for a load Z_L given by

$$r = \frac{Z_L - Z_0}{Z_L + Z_0} \qquad (10.1)$$

This reflection decreases the power delivered to the load by a factor $(1-|r|^2)$ and results in a frequency dependence of the signal delivered to the load. Indeed, with L the length of the coaxial cable and v_{el} the propagation speed along the cable, one finds for the voltage over the load,

$$|V_{load}| = \frac{V_{gen}}{2} \sqrt{1 + r^2 + 2\, r \cos\!\left(\frac{4\pi f}{V_{el}} L\right)} \qquad (10.2)$$

Since the differential resistance of a laser diode is typically around 5Ω and the electrical path between coaxial cable and laser diode typically has a characteristic resistance of 40Ω to 50Ω, there is also a reflection from the chip. This causes another decrease of the power delivered to the chip. However, the electrical path from coaxial cable to chip is usually much shorter than the coaxial cable and therefore the reflection from the chip does not generally cause a frequency dependence of the signal delivered to the chip. However, some care must be taken with long bond wires. They correspond to large inductances that can drastically reduce the bandwidth.

Planar technology, which allows photolithographic structuring, is generally preferred for optimum adaptation of connector, package, and chip. Some of the most used planar transmission lines are depicted in Figures 10.7 and 10.8 [14]. Microstrip lines (Figure 10.7) can handle high powers, require only a small area, and exhibit a low radiation loss, but their strip width is fixed for a given substrate and impedance. Coplanar waveguides (Figure 10.8) can handle less power, exhibit more radiation loss, and require a larger area than microstrip lines. However, a given substrate and impedance only define the ratio between strip width and gap width in these

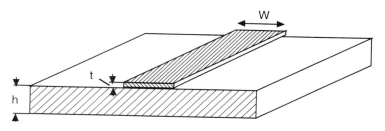

Figure 10.7 The microstrip for electrical connections. (*After:* [15].)

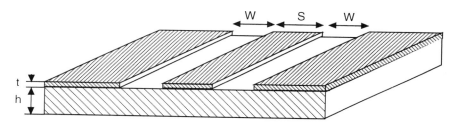

Figure 10.8 Coplanar waveguide for electrical connections. (*After:* [15].)

waveguides, and hence a geometrical adaptation of the strip width to connectors and to the chip is possible while maintaining a constant impedance.

10.3.2 Optical Aspects

Optical coupling is not straightforward, since the geometries of laser waveguide and optical fiber give rise to large differences in their respective mode profiles. Standard dimensions of single-mode fibers for the 1.3- to 1.55-μm wavelength range are a core diameter of 9 μm and a cladding diameter of 125 μm. For the small refractive index contrast of around 0.01 in optical fibers, one then typically finds a circular mode with a beam radius of 10 μm or more. The core of a laser waveguide is normally rectangular, with a width of a few micrometers in the lateral direction and a height of 0.1 to 0.2 μm in the transverse direction. For the typical refractive index contrast of 0.1 to 0.2 in the InGaAsP/InP system, one then finds an elliptical mode with a width of ±5 μm and a height of a few micrometers. The far field of the laser's mode profile is, in the transverse direction, distributed over a large angle, and a considerable part of it is emitted under angles that are larger than the numerical aperture of the single-mode fiber (Figure 10.9).

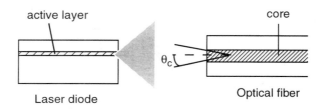

Figure 10.9 Far-field angle of the laser diode output light in relation to the numerical aperture of a single-mode optical fiber.

An approximate value for the coupling efficiency is easily obtained by calculating the overlap between the mode of the fiber and the mode of the laser diode, approximating both as Gaussian beams [15]. The mode of the fiber shows a cylindrical symmetry and can as a function of the distance r to the center of the fiber be expressed as

$$\psi_f = \sqrt{\frac{2}{\pi w_f^2}} \exp\left(-\frac{r^2}{w_f^2}\right) \qquad (10.3)$$

with the width w_f being approximately 1.1 times the core radius a. The mode of the laser waveguide has a different width in transverse (x) and lateral (y) directions and can be expressed as

$$\psi_l = \sqrt{\frac{2}{\pi w_x w_y}} \exp\left(-\frac{x^2}{w_x^2} - \frac{y^2}{w_y^2}\right) \qquad (10.4)$$

The power coupled from laser waveguide to fiber or vice versa is then given by the overlap integral T

$$T = \left| \int_{-\infty}^{+\infty} \int_{-\infty}^{+\infty} \psi_l(x, y) \psi_f^*(x, y) \, dx \, dy \right|^2 = 4 \left(\frac{w_f w_x}{w_f^2 + w_x^2} \right) \left(\frac{w_f w_y}{w_f^2 + w_y^2} \right) \qquad (10.5)$$

This overlap function is depicted in Figure 10.10 as a function of w_x/w_f for $w_x = w_y$. For a fiber mode with a width w_f of 5 μm and a laser mode with $w_x = 1$ μm and $w_y = 3$ μm, one finds a coupling loss of 4.7 dB.

Maximum overlap obviously results if waveguide and fiber are adjacent ($Z = 0$), aligned ($\theta = 0$), and centered ($d = 0$) (Figure 10.11). However, it is also important to know the decrease of the coupling efficiency with displacements (Z and d) and with tilt (θ). Indeed, large tolerances for displacements and tilt imply a less critical and

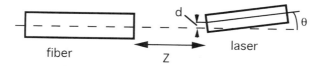

Figure 10.10 Different possible misalignments and tilts when light from a laser diode is coupled into a single-mode fiber.

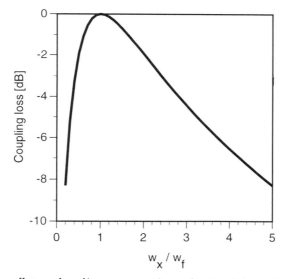

Figure 10.11 Coupling efficiency from fiber to waveguide as a function of the mode width ratio for $w_x = w_y$

faster alignment. The variation of the coupling efficiency as a function of the displacements and the tilt can be approximated again using Gaussian beams. For example, for a transverse misalignment d, the coupling efficiency can be calculated by replacing expression (10.4) for ψ_l with

$$\psi_l = \sqrt{\frac{2}{\pi w_x w_y}} \exp\left(-\frac{(x-d)^2}{w_x^2} - \frac{y^2}{w_y^2}\right) \tag{10.6}$$

and for the coupling efficiency one finds

$$T = T(d = 0) \exp\left(-\frac{2d^2}{w_f^2 + w_x^2}\right) \tag{10.7}$$

which is displayed in Figure 10.12. The influence of a lateral displacement can be calculated similarly using w_y instead of w_x. The influence of a longitudinal displacement Z can be derived using the transformation of the beam width w of a Gaussian beam due to diffraction. The influence of a tilt θ can be calculated by projecting the mode profile (10.3) over an angle θ.

The alignment tolerances and the maximum efficiency for coupling the laser light into a single-mode fiber are generally too small for practical applications. Several methods to overcome the small tolerances and efficiency do exist, however. The simplest method makes use of a single lens between laser and fiber. This allows the conversion of the laser mode profile to a beam profile with better overlap with the fiber mode. The method leads to good coupling efficiencies only when the laser mode is relatively circular (with $w_x = w_y$), as the lens is generally. Better results are usually obtained using combinations of lenses. One of these lenses can be a fiber lens or a lens formed by etching (or arc fusion rounding) of the fiber end. A somewhat more complex method is to integrate the laser diode with a waveguide taper. Different technolo-

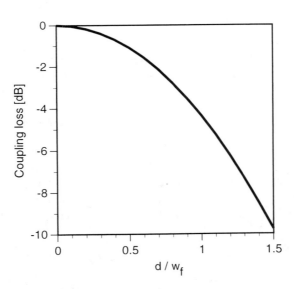

Figure 10.12 Coupling efficiency from fiber to waveguide as a function of the lateral misalignment for $w_f = w_y = w_x$.

gies can be used to realize the usually vertical tapering, such as selective-area MOVPE growth, shadow-masked MOVPE growth, temperature-dependent MBE growth. The majority of these technological processes, however, are still under development.

References

[1] Herman, M. A., and H. Sitter, *Molecular Beam Epitaxy: Fundamentals and Current Status*, Berlin: Springer-Verlag, 1989.
[2] Panish, M. B., and H. Temkin, *Gas-Source Molecular Beam Epitaxy*, Berlin: Springer-Verlag, 1993.
[3] Stringfellow, G. B., *Organometallic Vapor-Phase Epitaxy: Theory and Practice*, New York: Academic Press, 1989.
[4] Arai, S., M. Asada, T. Tanbun-Ek, Y. Suematsu, Y. Itaya, and K. Kishino, "1.6 μm Wavelength GaInAsP/InP Lasers," *IEEE J. Quant. Electron.*, Vol. 17, 1981, pp. 640–645.
[5] Adachi, S., and H. Kawaguchi, "Chemical Etching Characteristics of (001) InP," *J. Electrochem. Soc.*, Vol. 128, 1981, pp. 1342–1349.
[6] Tadakoro, T., F. Koyama, and K. Iga, "A Study on Etching Parameters of a Reactive Ion Beam Etch for GaAs and InP," *Jap. J. Appl. Phys.*, Vol. 27, 1988, pp. 389–392.
[7] Fonash, S. J., "Advances in Dry Etching Processes—A Review," *Solid State Technology*, January 1985, pp. 150–158.
[8] Panish, M. B., and S. Sumski, "Gas Source Molecular Beam Epitaxy of $Ga_xIn_{1-x}P_yAs_{1-y}$," *J. Appl. Phys.*, Vol. 50, 1984, pp. 3571–3576.
[9] Calawa, A. R., "On the Use of AsH_3 in the Molecular Beam Epitaxy of GaAs," *Appl. Phys. Lett.*, Vol. 38, 1981, pp. 701–703.
[10] Westbrook, L. D., A. W. Nelson, C. Dix, "High-Quality InP Surface Corrugations for 1.55μm InGaAsP DFB Lasers Fabricated Using Electron-Beam Lithography," *Electron. Lett.*, Vol. 18, 1982, pp. 863–865.
[11] Matthews, M. R., B. M. Macdonald, and K. R. Preston, "Optical Components—The New Challenge in Packaging," *IEEE Trans. Comp., Hybrids, Manu. Tech.*, Vol. 13, 1990, pp. 798–806.
[12] Khoe, G.-D., and H. G. Kock, "Laser-to-Monomode-Fiber Coupling and Encapsulation in a Modified TO-5 Package," *J. Lightwave Tech.*, Vol. 13, 1990, pp. 1315–1320.
[13] Reith, L. A., J. W. Mann, G. R. Lalk, R. R. Krchnavek, N. C. Andreadakis, and C. Zah, "Relaxed-Tolerance Optoelectronic Device Packaging," *J. Lightwave Tech.*, Vol. 9, 1990, pp. 477–484.
[14] Gupta, K. C., R. Garg, and I. J. Bahl, *Microstrip Lines and Slotlines*, Norwood, MA: Artech House, 1992.
[15] Kogelnik, H., "Matching of Optical Modes," *Bell Systems Tech. J.*, Vol. 43, January 1964, pp. 334–337.

Chapter 11

Epilogue

11.1 TRENDS IN OPTICAL TRANSMISSION AND THE IMPACT ON DFB LASERS

The trend in optical transmission systems is for higher bit rates, higher powers, and longer span lengths. Moreover, the advent of rare-earth-doped fiber amplifiers has changed the outlook of optical transmission systems considerably. Even before optical amplifiers were in commercial use, long-distance transmission became focused on the 1,550-nm wavelength window because the fiber attenuation is minimal in this window. However, in standard fiber, the dispersion is considerably higher in the 1,550-nm window (≈ 17 ps/km-nm) than in the 1,300-nm window (≈ 0.1 ps/km-nm). As a result, significant dispersion penalties may be incurred in the 1,550-nm window if no measures are taken.

A first measure consists of using chirp-free lasers. Therefore, directly modulated DFB lasers with new strained MQW structures are being investigated for minimal chirp (with α as low as 1). In non-return-to-zero (NRZ)-modulated signals, the majority of chirp occurs during the rise and fall transients. For lasers with extremely high −3-dB modulation bandwidths (e.g., over 25 GHz), very short rise and fall transients will be possible if the laser is modulated at about 10 Gbps. A narrowband optical filter behind the laser can then be used to remove the transient chirp (≈ 0.2 nm) from the output of the laser.

A second way to obtain a chirp-free source is the use of DFB lasers and external modulators. Especially for high bit rates at and above 2.5 Gbps, a lot of research is being conducted on monolithically integrated DFB lasers and electro-absorption (EA) modulators. These devices are usually based on MQW structures, where the MQW structure can be engineered differently for the laser and the EA sections. For such DFB-EA laser diodes, the modulation speed is basically determined by a resistive-capacitive (RC) time constant. If the device capacitance can be made small enough, it becomes possible to obtain sources with extremely low chirp, with speeds over 10 Gbps, and without the complication of the relaxation resonance, occurring in directly modulated lasers.

Nevertheless, even for chirp-free sources, the dispersion limit in standard fiber for 10 Gbps corresponds to a maximum distance of approximately 70 km for a 1-dB power penalty in the absence of fiber nonlinearities. To increase this distance limitation, the dispersion coefficient must be decreased or nonlinear transmission techniques must be used.

The amplifier-free transmission distance is not only bound by the dispersion limit but also by fiber attenuation and fiber nonlinearities. In order to obtain large repeater spacings erbium-doped fiber amplifiers (EDFA) can be used to boost the signal powers to very high levels. However, with such high signal powers, optical nonlinearities (such as self-phase modulation, cross-phase modulation, four-photon mixing, stimulated Raman scattering, and stimulated Brillouin scattering) will occur and they can greatly reduce the performance of the fiber transmission system. In practice, the system performance is determined by the interaction of source chirp, source power, amplifier characteristics, amplifier spacing, fiber nonlinearities, and the fiber chromatic dispersion.

In standard single-mode fiber, transmission beyond the dispersion limit has been achieved in different ways [1]: soliton transmission, optical phase conjugation, dispersion compensation, duobinary signaling, dispersion-supported transmission (DST), and prechirp transmission. The latter two techniques are based on combined AM and FM. In both techniques, a controlled interaction between the source chirp, self-phase modulation, and dispersion is used to increase the transmission distance in unrepeatered systems (i.e., systems without inline optical amplifiers). However, in DST, the combined modulation is introduced by direct modulation of the transmitter laser, while in prechirp transmission, an EA modulator is used.

In dispersion-shifted fiber (DSF), the zero-dispersion wavelength is shifted into the 1,550-nm window. For dispersions below 0.1 ps/km-nm, the dispersion-related maximum distance for chirp-free sources at 10 Gbps is now close to 10,000 km. However, to match the wavelength of the laser source with the zero-dispersion wavelength of the DSF, the laser wavelength must be specified within a tight interval [2]. Moreover, fiber nonlinearities will often limit the system performance to distances well below the dispersion limit.

In practice, nonlinearities can be countered by having a local dispersion that is nonzero. If, at the same time, a link-averaged zero dispersion is maintained, the nonlinearities and the dispersion can interact to minimize both effects. Such dispersion management techniques are currently under investigation for long-distance transmission.

All the above systems are based on time-domain-multiplexed (TDM) systems, in which the source delivers the high-bit-rate modulated signal. The maximum bit rate of such systems is limited by the maximum bandwidth of the source. With current technology, this maximum electrical modulation rate is believed to be in the range of 20 to 40 Gbps. However, considerable research is still required to reach this 40-Gbps objective. The picture at present is that directly digitally modulated lasers will not provide 40 Gbps. Modulators (both semiconductor and fiber types) can achieve 40 Gbps, and efficient detectors are also within reach of 40 Gbps.

Exploiting the fiber capacity beyond 40 Gbps requires other techniques such as optical time-domain multiplexing (OTDM) or wavelength-division multiplexing (WDM) [3]. The latter approach also offers an alternative way to extend transmission beyond the dispersion limit. Currently, several experimental (8×5 Gbps over 8,000 km; 17×20 Gbps over 150 km; 16×10 Gbps over 1,000 km) and even some field WDM systems (4×2.5 Gbps over 3,700 km; 3×5 Gbps over 2,100 km) have been reported [4]. For short distances, even 1-Tbps WDM systems have already been achieved (55×20 Gbps over 150 km [5], 50×20 Gbps over 55 km [6]; 10×100 Gbps over 40 km [7]).

If the WDM channels are modulated at such high rates, dispersion also becomes a problem for each of the channels. Therefore, dispersion management techniques applied to long spans must take into account the different channels in the link design.

WDM systems require multiple sources with closely spaced emission wavelengths. Multiwavelength DFB laser arrays form a potential solution. Key issues here are a high yield for the arrays, some fine tunability of the individual wavelengths, and long-term wavelength stability. The tunability can, for instance, be useful to obtain unequal channel spacings in order to avoid four-photon mixing. A high yield of single-mode operation at predictable Bragg modes for the individual array elements can be achieved through the use of gain coupling [8]. These arrays could also be extended to carry the EA modulators associated with the lasers, an optical combiner, and possibly also a semiconductor optical amplifier as power booster. Although such photonic integrated circuits (PIC) may show increased chip costs, this will be offset by reduced packaging costs.

With the introduction of WDM transmission, the concept of wavelength routing and switching has also emerged. Wavelength switching allows wavelength reuse and reduces wavelength blocking, but it requires wavelength converters. Ideally, a wavelength conversion should be transparent to the bit rate and signal format. Wavelength conversion can be achieved in different ways [9]. One promising way is based on the use of cross-gain or cross-frequency modulation of laser diodes. This approach is simple but lacks signal format transparency. Moreover, considerable work is still required to reduce the signal degradation and realize the cascadability of many converters.

Up to now we have only discussed the future evolution of high-bit-rate long-span digital communication systems and the role DFB lasers could play there. However, optical communications is also rapidly expanding across the access network. Digital broadband access networks (e.g., fiber-in-the-loop (FITL) networks with an aggregate downstream data rate of 622 Mbps for 16 downstream optical network terminations and 2 Mbps upstream rate per optical network termination) can still be operated by Fabry-Perot lasers. In the near future, however, DFB lasers will see a considerable cost reduction and may then start to replace Fabry-Perot devices. This will become even more imperative if the aggregate data rate of these broadband systems goes up (e.g., with the advent of super passive optical networks with up to 1,000 downstream optical terminations).

Besides digital communications, analog communications are also important here. Two main analog applications can be distinguished: CATV feeder networks (with, for instance, subcarrier multiplexing) and antenna feeders for wireless cellular networks (carrying broadband microwave signals potentially up to 40 GHz). For CATV systems, the current trend in research is toward higher numbers of channels (for instance, up to 500 channels, 6 MHz/channel). Such applications require DFB lasers with a wide modulation bandwidth, a narrow linewidth, very low distortion characteristics, and low chirp. Moreover, high-output-power lasers are preferred in order to avoid the use of EDFAs, because they increase the cost of the access. Simple, directly modulated DFB lasers also provide higher coupled output power than more complex devices with an integrated modulator. However, while direct modulation of microwave signals offers considerable modulation bandwidth (demonstrated up to about 40 GHz), together with small size and simplicity, external modulators offer better linearity, higher modulation bandwidths, narrower linewidths, and lower noise. On the other hand, external modulation costs more, shows polarization dependence, and requires higher drive powers.

For access networks, where the cost per subscriber needs to be sufficiently low to justify the investment, DFB lasers are currently only justifiable in the downstream direction. For CATV broadcast systems, which mainly operate in the downstream direction, this is no problem. However, for cellular antenna feeders that need to carry a more symmetric load in both directions, DFB lasers are required at both ends.

Finally, note that multisection DFB lasers with inhomogeneous injection may offer interesting functionality in photonic signal processing as self-pulsating clock regenerators or as bistable switches.

In conclusion, we see that the evolution in optical transmission and network technology continues to create opportunities for the usage of DFB lasers. Basically, these lasers will be used as the preferred single-mode, narrow-linewidth source for high-bit-rate optical transmission systems. The modulation of their output will either be direct or via external modulators. External modulators offer lower chirp and they will increasingly be integrated on the laser chip. Low-chirp lasers can also be obtained by carefully designing the laser, especially when using MQW structures. Moreover, some controlled chirp is sometimes desired to reduce the effects of fiber nonlinearities. Careful engineering of the source chirp, therefore, becomes an important issue.

11.2 FUTURE DIRECTIONS IN DESIGN AND MANUFACTURING OF DFB LASERS

At the time of writing, the fabrication of DFB lasers is slowly but definitely moving from a research activity into a mature production activity. Indeed, the intense research in the past years has led to extensive knowledge about the designs required to achieve a priori specified characteristics, about the optimization of fabrication and process reliability, and about the packaging. Current research activities are therefore aimed more and more at increasing the fabrication efficiency and yield, or, in other words, reducing the manufacturing costs.

One major factor that is still hindering the manufacturing costs is the unclear relationship between the process (or growth) parameters and the physical parameters of the grown component. Fabricating a component that corresponds with the intended design therefore requires many cycles of fabrication and characterization. This particularly increases costs when components for specific applications are to be produced in rather small volumes. A second major drawback is that commercially available laser diodes are generally described by a limited set of measured characteristics, a fact that, for example, prevents the evaluation of the suitability of standard components for use in one or another system application using system simulation software. To improve the current situation, extensive research is conducted on parameter extraction from measurements on laser diodes. An example of such parameter extraction from measured subthreshold spectra was briefly introduced in Chapter 6. Several other examples of the parameter extraction from isolated measurements have been reported in the literature [10–13]. The final goal of this research is, however, to be able to extract a complete and reliable set of laser parameters from a variety of measurements. This requires that yet other parameter extraction possibilities and the consistency of the extracted parameters with all possible measured characteristics be explored.

A second factor that has a major impact on the price of packaged laser modules is the complexity of the packaging and especially the manpower required to align laser diode, lenses, isolator, and fiber pigtail. This is mainly caused by the difference in the mode profile of the laser and fiber, as was explained in Chapter 10. The problem is tackled in two ways. A first approach is to integrate the laser with a tapered waveguide in order to adapt the mode profile at the facet of the chip to the mode profile of the fiber. A second approach is an attempt to fully automate the pigtailing. This last approach will obviously be more advantageous (i.e., cheaper), since it eliminates any human interference.

A third factor that is of key interest in reducing the cost and improving the reliability of laser diodes is the possibility that laser diodes will operate over a wide temperature range without a Peltier cooler. The wide temperature operation also implies that high optical powers can be obtained with low driving currents and that the external laser characteristics will change little with temperature. This requires the use of optimized strained MQW structures, which should make DSM operation feasible in the -40 to $+125°C$ temperature range. Simultaneously, the laser lifetime will become longer.

A last remark that cannot be omitted when discussing fabrication efficiency is that gain-coupled (or rather partly gain-coupled) DFB lasers exhibit a larger single-mode yield than index-coupled lasers. Since single-mode behavior is generally one of the basic requirements, this larger single-mode yield also implies a larger fabrication yield. The main reasons for the higher costs of DFB lasers compared to Fabry-Perot lasers are the yield of single-mode lasers and, secondly, the higher effort required for selection and characterization. Partly gain-coupled lasers should therefore be preferred when the fabrication costs are considered. For index-coupled lasers (AR/HR-coated), yields of only 40% to 50% are feasible, whereas for gain-coupled DFB lasers (AR/AR-coated), yields of 80% to 90% are possible. At present, more and more companies are indeed including gain-coupled structures in their portfolio of laser diodes. However, the vast majority of fabricated devices are still purely index-coupled lasers. A reason is that the fabrication of partly gain-coupled lasers is a little more complex than that of index-coupled lasers. Starting the fabrication of partly gain-coupled lasers thus requires additional development costs.

Finally, it is worth mentioning that DFB fiber lasers have also recently appeared. These lasers can be directly spliced to the transmission fibers, and the introduction of a $\lambda/4$ phase shift guarantees single-mode operation. Much of the basic theory on DFB lasers treated in this book (mainly Chapters 3 and 4) can be applied to such fiber lasers. Although they require external modulation and are considerably less compact than DFB laser diodes, DFB fiber lasers may become a valid alternative in some cases.

References

[1] Jorgensen, B. F., R. J. S. Pedersen, and C. J. Rasmussen, "Transmission of 10 Gbit/s Beyond the Dispersion Limit of Standard Single Mode Fibers," *Proc. 21st Eur. Conference on Opt. Comm. (ECOC'95)*, Brussels, 17–21 September 1995, pp. 557–564.

[2] Garthe, D., W. S. Lee, R. A. Saunders, and A. Hadjifotiou, "20 Gbit/s TDM Transmission Over 560 km of Dispersion Shifted Fibre With a Transmission Window of More Than 12 nm," *Proc. 21st Eur. Conference on Opt. Comm. (ECOC'95)*, Brussels, 17–21 September 1995, pp. 597–600.

[3] Mahony, M. J., "Optical Multiplexing in Fiber Networks: Progress in WDM and OTDM," *IEEE Communications Magazine*, Vol. 33, No. 12, December 1995, pp. 82–88.

[4] Gnauck, A. H., "Recent Progress in High-Capacity Long-Haul WDM Systems," *Optical Fiber Communication (OFC'96)*, Technical Digest, Vol. 2, San Jose, 25 February–1 March, pp. 47–48.

[5] Onaka, H., H. Miyata, G. Ishikawa, K. Otsuka, H. Ooi, Y. Kai, S. Kinoshita, M. Seino, H. Nishimoto, and T. Chikama, "1.1 Tbps WDM Transmission Over a 150 km 1.3 μm Zero-Dispersion Single Mode Fibre," *Optical Fiber Communication (OFC'96)*, Post Deadline Papers, Part B, San Jose, 25 February–1 March, pp. PD19-1/5.

[6] Gnauck, A. H., A. R. Chraplyvy, R. W. Tkach, J. L. Zyskind, J. W. Sulhoff, A. J. Lucero, Y. Sun, R. M. Jopson, F. Forghieri, R. M. Derosier, C. Wolf, and A. R. McCormick, "One Terabit/s Transmission Experiment," *Optical Fiber Communication (OFC'96)*, Post Deadline Papers, Part B, San Jose, 25 February–1 March, pp. PD20-1/5.

[7] Morioka, T., H. Takara, S. Kawanishi, O. Kamatani, K. Takiguchi, K. Uchiyama, M. Saruwatari, H. Takahashi, M. Yamada, T. Kanamori, and H. Ono, "100 Gbps x 10 Channel OTDM/WDM Transmission Using a Single Supercontinuum WDM Source," *Optical Fiber Communication (OFC'96)*, Post Deadline Papers, Part B, San Jose, 25 February–1 March, pp. PD21-1/5.

[8] Makino, T., G. P. Li, A. Saranga, and W. Huang, "Multiwavelength Gain-Coupled MQW DFB Laser

Array With Fine Tunability," *Optical Fiber Communication (OFC'96)*, Technical Digest, Vol. 2, San Jose, 25 February–1 March, pp. 298–299.

[9] Mikkelsen, B., T. Durhuus, C. Joergensen, S. L. Danielsen, R. J. S. Pedersen, and K. E. Stubkjaer, "Wavelength Conversion Devices," *Optical Fiber Communication (OFC'96)*, Technical Digest, Vol. 2, San Jose, 25 February–1 March, pp. 121–122.

[10] Morton, P. A., D. A. Ackerman, G. E. Shtengel, R. F. Kazarinov, M. S. Hybertsen, T. Tanbun-Ek, R. A. Logan, and A. M. Sergent, "Gain Characteristics of 1.55 µm High-Speed Multiple-Quantum-Well Lasers," *IEEE Phot. Tech. Lett.*, Vol. 7, 1995, pp. 833–835.

[11] Joindot, I., and J. L. Beylat, "Intervalence Band Absorption Coefficient Measurements in Bulk Layer, Strained and Unstrained Multiquantum Well 1.55µm Semiconductor Lasers," *Electron. Lett.*, Vol. 29, 1993, pp. 604–606.

[12] Su, C., V. Lanzisera, and R. Olshansky, "Measurement of Non-Linear Gain From FM Modulation Index of InGaAsP Lasers," *Electron. Lett.*, Vol. 21, 1989, pp. 893–895.

[13] Westbrook, L., "Measurement of dg/dN and dn/dN and Their Dependence on Photon Energy in $\lambda = 1.55$ µm InGaAsP Laser Diodes," *IEE Proc. Pt. J*, Vol. 133, 1986, pp. 135–142.

Glossary

AR	antireflection
ASE	amplified spontaneous emission
BH	buried heterostructure
C^3	cleaved coupled cavity
CATV	cable television
CLADISS	compound cavity laser diode simulation software
CMBH	capped-mesa buried heterostructure
CNR	carier-to-noise ratio
CSO	composite second order
CTB	composite triple beat
CW	continuous wave
DBR	distributed Bragg reflector
DCPBH	double-channel planar buried heterostructure
DFB	distributed feedback
DH	double heterostructure
DSF	dispersion-shifted fiber
DSM	dynamic single mode
DST	dispersion-supported transmission
EA	electro-absorption
EDFA	erbium-doped fiber amplifier
EMBH	etched-mesa buried heterostructure
FP	Fabry-Perot
FFT	fast Fourier transform
FITL	fiber-in-the-loop
FM	frequency modulation
FPI	Fabry-Perot interferometer
FSR	free spectral range
GaAs	gallium arsenide
GSMBE	gas source MBE
IM	intensity modulation
IVBA	intervalence band absorption

LPE	liquid-phase epitaxy
MBE	molecular beam epitaxy
MI	Michelson interferometer
MOCVD	metal-organic chemical vapor deposition
MOVPE	metal-organic vapor-phase epitaxy
MQW	multiquantum well
NR	Newton-Raphson
OEIC	optoelectronic integrated circuits
OMD	optical modulation depth
OTDM	optical time-domain multiplexing
PIC	photonic integrated circuit
RIN	relative intensity noise
SCH	separate confinement heterostructure
SCM	subcarrier multiplexing
SDH	synchronous digital hierarchy
SE	stimulated emission
SI-InP	semi-insulating InP
SIBH	semi-insulating buried heterostructure
SL-MQW	strained-layer multiquantum well
SMSR	side-mode suppression ratio
SNR	signal-to-noise ratio
SQW	single quantum well
TBA	tertiary butyl arsine
TDM	time-domain-multiplexing
TE	transverse electric
TM	transverse magnetic
TTG	tunable twin guide
UHV	ultrahigh vacuum
VPE	vapor-phase epitaxy
WDM	wavelength-division multiplexing

List of Symbols

L	laser length [μm]
w	width of active layer [μm]
d	thickness of active layer [μm]
$V_{act} = wdL$	volume of active layer [μm^3]
x	coordinate in the transverse direction [μm]
y	coordinate in the lateral direction [μm]
z	coordinate in the longitudinal direction [μm]
λ	wavelength [μm]
$k = 2\pi/\lambda$	wave number [μm^{-1}]
E	electric field [V/μm]
Γ	confinement factor
n_e	effective refractive index
v_g	group velocity [cm/s]
Λ	grating period [μm]
$\lambda_\beta = 2n_e\Lambda$	Bragg wavelength [μm]
κ	coupling coefficient [μm^{-1}]
κ_n	index-coupling coefficient [μm^{-1}]
κ_g	gain-coupling coefficient [μm^{-1}]
g	power gain of the material [μm^{-1}]
$G = \Gamma g v_g$	modal gain [sec^{-1}]
dG/dN	differential modal gain [cm^3/s]
ξ	gain compression coefficient [cm^3]
τ_d	differential carrier lifetime [sec]
R_{sp}	spontaneous emission rate [sec^{-1}]
n_{sp}	inversion factor
α_{int}	internal power loss coefficient [μm^{-1}]
$2\alpha_{end}$	facet loss [μm^{-1}]
$\gamma = v_g(\alpha_{int} + 2\alpha_{end})$	modal loss [sec^{-1}]
N	carrier (electron) density [μm^{-3}]
S	photon density [μm^{-3}]
A	monomolecular recombination rate [1/s]

Symbol	Description
B	bimolecular recombination rate [$\mu m^3/s$]
C	Auger recombination rate [$\mu m^6/s$]
P	optical power [mW]
$r_{1,2}$	field reflection coefficient at the facets
$R_{1,2}$	power reflection coefficient at the facets
R_s	series resistance of the cladding layers [Ω]
V_{DH}	voltage over the heterojunction [V]
E_g	bandgap energy [eV]
F_n	quasi-Fermi level of electrons [eV]
F_p	quasi-Fermi level of holes [eV]
I	current [mA]
J	current density [mA/μm^2]
q	electron charge [C]
D_n	diffusion coefficient of electrons [cm^2/s]
D_p	diffusion coefficient of holes [cm^2/s]
μ_n	electron mobility [cm^2/Vs]
μ_p	hole mobility [cm^2/Vs]
L_n	diffusion length of electrons [μm]
L_p	diffusion length of holes [μm]
ρ	resistivity [$\Omega\,\mu m$]
R_s	series resistance [Ω]
T	temperature [K]
kT	thermal energy [eV], k: Boltzmann constant
N_A	acceptor doping concentration [cm^{-3}]
N_D	donor doping concentration [cm^{-3}]
m	(optical) modulation depth
Ω	modulation pulsation [Hz]
ω	optical pulsation [Hz]
ν	optical frequency [Hz]
h	Planck constant
θ	damping constant of the relaxation oscillations [Hz]
Ω_r	resonance pulsation of the relaxation oscillations [Hz]
$\Delta\beta$	complex Bragg deviation [μm^{-1}]
$\Delta\beta_i$	imaginary part of the Bragg deviation [μm^{-1}]
$\Delta\beta_r$	real part of the Bragg deviation [μm^{-1}]
s	number of photons per unit length [μm^{-1}]
s^+	number of forward propagating photons/unit length [μm^{-1}]
s^-	number of backward propagating photons/unit length [μm^{-1}]
α	linewidth enhancement factor
α_{eff}	effective linewidth enhancement factor
$S_X(f)$	spectral density of X at frequency f

About the Authors

Geert Morthier is researcher and group leader at the department of Information Technology, University of Gent—IMEC. He has conducted research on DFB lasers for nearly ten years and has published over fifty technical articles and papers on this topic. He has contributed significantly to the understanding of harmonic distortion and linewidth rebroadening, among other topics.

Dr. Morthier holds a MScEE and a Ph.D. in electrical engineering, both from the University of Gent. He is also a member of the IEEE and OSA.

Patrick Vankwikelberge has been involved in research on telecommunications for nearly 10 years, first with the Department of Information Technology, University of Gent-IMEC, where he worked on DFB lasers, and then with the Alcatel Corporate Research Center in Antwerp, where he was involved in research on broadband networks. After a brief period with the RACE Central Office at the European Commission and with the SIDMAR Corporate Network Department, he recently joined Alcatel's Corporate Business Processes and Information Systems Department in Brussels to work on process engineering.

Dr. Vankwikelberge holds a MScEE and a Ph.D. in electrical engineering, both from the University of Gent. He is also a member of the IEEE.

Index

1/4 phase-shifted DFB lasers, 102–4
  phase resonance at Bragg wavelength, 102
  second-order distortion in, 230, 231
  third-order distortion in, 231
  threshold resonance solutions, 103
  See also DFB lasers
Absorption
  carrier, 37
  interband, 84
  intervalence band (IVBA), 38
Amplified spontaneous emission (ASE), 148–55
  in side mode, 147
  spectrum measurement, 179–81
AM response
  frequency dependence of, 49
  obtaining, 46
  static, 50
  See also FM response
Antireflection (AR)-coated DFB lasers, 93
  AR coating application, 247
  laser resonance absence in, 97
  threshold resonance condition, 96
  See also DFB lasers
Avalanche photodiode (APD), distortion in, 220

Beam propagation method, 118
BH laser diode, index profile, 66
Boundary conditions, 88–90
  for electromagnetic fields, 89
  expression of, 88
Bragg reflection, 76
  characteristic, 24
  distributed, 203
Bragg wavelengths, 76, 97, 170
Bragg windows, 76, 77
Broadband approach, 113–14
  advantages, 114
  laser equations, 114
  See also Coupled-mode model;
    Coupled-mode theory
Bulk materials, 38–39
  gain, 38–39
  refractive index, 38–39, 40
Buried heterostructure (BH) lasers, 9

Capacitances, 123, 140
  depletion, 140, 141
  differential, 140
  diffusion, 123
  distributed, 141
  junction, 141–42
  minimum of upper junction, 142
  See also Resistances
Capped-mesa buried heterostructure
    (CMBH), 239
Carrier absorption, 37
Carrier density
  active layer, 133
  average, 160
  as continuous quantity, 84
  dependence, 56
  fluctuations in, 256–57, 263
  linear function of, 54
  nonuniform, 162, 227, 234
  threshold, 54
  variation during switch-on/switch-off, 53
  variation of loss, 266
Carrier density rate equations, 30–31
  electrical transport problem and, 83–85
  overview, 45
  See also Rate equations
Carrier heating, 43–44

Carrier injection, 123–45
  bottleneck, 214
  in gain-guided lasers, 134–36
  in index-guided lasers, 134–36
  in oxide stripe laser, 134
Carrier leakage, 124
  circuit modeling, 144–45
  in index-guided structures, 136–40
  over heterobarriers, 131–33
  parts of, 131
Carrier-to-noise ratio (CNR), 20
CATV
  broadcast systems, 296
  feeder networks, 296
  second-order distortion and, 248
Characteristic equation, 93
Chirp
  adiabatic, 199
  characterization using fiber dispersion, 192–93
  defined, 18
Chirped-grating lasers, 170
  fabrication of, 171
  pure, 172
Cladding layers, 123
  P-cladding layer, 131
  resistance, 141
  thickness, 132
CLADISS computer model, 107
Clipping distortion
  analog system with, 243
  prevention of, 244
  rms modulation index vs., 245
  See also Harmonic distortion
Clipping effects, 243
Complex-coupled DFB lasers, 25, 116
Composite second-order (CSO), 220
  from clipping, 244
  defined, 241
  obtaining, 243
  plotted, 244, 245
Composite triple beat (CTB), 220
  from clipping, 244
  defined, 241
  obtaining, 243
Continuous tuning, 177
Continuous-wave analysis, 109–11
  defined, 109
  purpose of, 109
  See also Narrowband approach
Coplanar waveguide, 287
Corrugated-pitch-modulated lasers, 170, 171
Corrugated waveguide structure, 118
Coupled-cavity lasers, 22–23

Coupled-mode model
  broadband approach, 113–14
  narrowband approach, 106–13
  numerical solutions, 105–6
Coupled-mode theory, 61–90
  applying, 93–119
  extended, 61–62
  longitudinal, 61
Coupled-wave equations
  Bragg wavelengths, 76
  for continuous wave (CW), 82
  coupling coefficients, 72
  discussions of, 75–83
  instantaneous optical frequencies, 81
  longitudinal rate equations for optical field, 79–80
  maximum gain wavelengths, 75–76
  radiation modes in higher order gratings, 76–79
  reduction toward, 70–75
  spontaneous emission, 81–83
Coupling coefficients, 93
  calculating, 118
  for DFB lasers, 117–19
  as function of tooth height, 119
Current-driven lasers, 85–86
Cutoff frequency, 272
  resistance and, 141
  spatial hole burning, 195–96, 235
  thermal, 199

Depletion capacitances, 123
Detuning, 274
DFB laser diodes
  coupled-mode theory, 61–90
  fabrication of, 279–91
  FM response measurement, 189–93
  harmonic distortion in, 219–33
  historical background, 2–4
  IM response measurement, 187–88
  intermodulation distortion in, 219–33
  noise characteristics, 251–75
  overview, xiii
  packaging of, 285–91
  spectrum, 147–84
  stripe contact, 9
DFB lasers
  l/4 phase-shifted, 102–4, 230–31
  AR-coated, 93, 96, 97
  ASE spectrum measurement, 179–82
  axial power variation in, 165
  badly designed, 77
  complex-coupled, 25, 116
  costs of, 297–98

Index

coupling coefficients, 117–19
DFB lasing modes, 76
Fabry-Perot lasing modes, 76
fiber, 298
FM noise, 254–59
FM response of, 200–210
future directions, 296–98
gain-coupled, 4, 25, 97–99, 173–77
grating, 9, 25
highly coherent, designing, 273–75
highly linear, designing, 245–48
high-speed, designing, 215–16
high-speed IM response, 196–98
impact of, 293–96
IM response, 194–95
index-coupled, 25, 95–97, 104–5, 169–73
linewidth, 259–61
multi-phase-shift, 4
multi-quantum-well (MQW), 4, 214–15, 225
multisection, 4, 17, 108, 296
phase-shifted, 93
with reduced spatial hole burning, 167–77
with reflecting facets, 101–2
RIN, 269–73
round-trip gain at threshold, 156
side-mode rejection, 155–58
single-mode yield, 157
structures of, 4
threshold optical spectrum, 155
threshold solutions, 94–105
wavelength tunability of, 17, 177–79
yield, 155–58
Differential gain, 39, 163, 235
Differential resistance, 286
Diffraction grating-based monochromator, 180
Diffusion capacitances, 123
Diffusion equation, 132
Diffusion voltage, 124
Discrete tuning, 177
Dispersion-shifted fiber (DSF), 294
Dispersion-shifted transmission (DST), 294
Distortion. *See* Clipping distortion;
    Intermodulation distortion
Distributed Bragg reflector (DBR) lasers, 4, 25
Distributed feedback lasers. *See* DFB laser
    diodes; DFB lasers
Double-channel planar buried heterostructure
    (DCPBH) laser, 123
equivalent circuit for, 124
illustrated, 137
Double heterojunctions
charge density and electric field, 126
formed by active layer, 132
voltage over, 167

Double heterostructure (DH) laser diodes, 3, 5
energy band diagram, 6
Fabry-Perot, 4
III-V compound semiconductors, 7
laser cavity, 5
Dynamic single-mode (DSM) laser
    diodes, 21–25
coupled-cavity lasers, 22–23
with distributed optical feedback, 23–25
injection-locked lasers, 23
short-cavity lasers, 21–22
structure types, 22

Electrical process, 62
Electrical transport problem, 83
carrier rate equation, 83–85
current/voltage drive, 85–86
Electro-absorption (EA) structure, 293
Electro-optic process, 62–63
Energy density, 75
Energy gap, 100–101
Equilibrium distribution
disturbance of, 43
thermal, 42
Erbium-doped fiber amplifiers (EDFA), 294
Etched-mesa buried heterostructure
    (EMBH), 136
illustrated, 137
leakage currents, 136
leakage paths, 137
P-layer doping level, 139
P-layer thickness, 139
External reflections, 58

Fabrication, 279–91
grating techniques, 284–85
LPE, 281–82
MBE, 282
MOVPE, 283–84
process steps, 280
See also Packaging
Fabry-Perot (F-P) laser diodes, 2
concept of, 10–12
DH, 4
optical spectrum of, 16
Fabry-Perot interferometers (FPIs), 180, 189–90
components of, 180
as filters, 289
FM measurements based on, 189–90
measurement analysis, 190
modulation frequency vs. resolution, 189
Fabry-Perot lasers
calculated optical spectrum of, 150, 151
concept of, 5

Fabry-Perot lasers (continued)
    FM response of, 198–200
    high-speed IM response, 196–98
    IM response, 194
    output power vs. injection current, 152, 153
    radiation modes, 76
    round-trip gain, 151
    second-order distortion in, 229
    spatial hole burning in, 228
    system view, 10
Fermi-Dirac functions, 39
Fiber-optic communication, 2
FM noise, 55
    calculation methods, 255–59
    in DFB lasers, 254–59
    frequency dependence, 254–55
    measuring, 252
    spectral density, 56, 257
    spectrum, 253
        finding, 256
        of semiconductor laser, 256
    See also Noise
FM response, 198–210
    of 1/4 phase-shifted DFB lasers, 209–10
    amplitude and phase
        shift, 200–201, 205–12
    calculation, 204, 206
    of DFB lasers, 200–210
    effective index modulation, 201
    of Fabry-Perot lasers, 198–200
    frequency dependence of, 48
    IM response relationship, 192
    at low/intermediate modulation
        frequencies, 207
    obtaining, 46
    static, 50
    See also IM response
FM response measurements, 189–93
    Fabry-Perot interferometers and, 189–90
    gated, delayed self-homodyne
        technique, 190–91
    laser chirp with fiber dispersion, 192–93
Free spectral range, 180
Frequency modulation (FM). See FM noise;
        FM response

Gain
    in bulk materials, 38–39
    clamping, 11, 12
    differential, 39, 163, 235
    dispersion implications, 269
    as function of photon energy, 40
    grating, 87
    material, 37

    maximum wavelengths, 75–76
    modal, 86, 99
    optical, 36–44
    oscillations, 18
    round-trip, 108, 151, 154, 156
    temperature dependence of, 14
    See also Gain suppression
Gain-coupled DFB lasers, 4, 97–99, 173–77
    with corrugated active layer, 98
    defined, 25
    fabrication of, 173
    first-order, 97
    with gain grating, 87
    with loss grating, 88
    standing-wave effect in, 86–88
    yield, 157, 158
    See also DFB lasers
Gain-guided lasers, 7
    carrier injection in, 134–36
    illustrated, 8
Gain suppression
    facet reflectivities, 228
    FM, 209
    gain value and, 224
    influence on harmonic distortion, 224–26
    linewidth and, 264–66
    See also Gain
Gas source MBE (GSMBE), 282
Grating fabrication, 284–85
    holographic exposure technique, 284
    in photoresist, 285
Gratings
    by double exposure technique, 174
    chirped, 171
    DBR laser, 25
    defined, 9
    DFB laser, 9, 25
    first-order, 78
    gain, 87
    high order, radiation mode influence, 76–79
    index, 73, 79
    loss, 88, 99
    modal gain, 99
    quasiperiodic, 73
    rectangular with variable duty cycle, 173
    reflections, 156
    second-order, 77, 78, 104
    third-order, 77, 78
    uncertainty of phase of, 89
    See also Grating fabrication
Green's functions, 258, 259
Group velocity, 73

Harmonic distortion, 50–52, 219–48

in APD, 220
caused by intrinsic nonlinearities, 222–23
cause of, 50
clipping-induced, 243
differential gain influence, 235
of electrical signal generator, 220
gain suppression influence, 224–26
leakage current influence, 237–40
low, designing for, 245–48
measuring, 220–21
relaxation oscillation influence, 221–24
second-order, 51, 221
spatial hole burning influence, 226–36
third-order, 52, 221, 231
Helmholz equation, 258
Henry's formula, 58
Heterojunctions, 124–28
double, 126, 132
EMBH, 136
SIBH, 138–40
High-gain approximation, 95
Highly coherent DFB lasers, 273–75
Highly linear DFB lasers, 245–48
High reflection (HR) coatings, 247
High-speed DFB lasers, 215–16
Holographic exposure technique, 284
IM response, 193–98
of l/4 phase-shifted DFB lasers, 209–10
amplitude as function of modulation
frequency, 195
angular cutoff bandwidth, 197–98
damping rate, 198
for DFB lasers, 194–95
of Fabry-Perot laser, 194
FM response relationship, 192
high-frequency, 196–98
maximum modulation bandwidth, 198
measuring, 187–88
spatial hole burning cutoff
frequency, 195–96
subgigahertz, 193–95
test setup, 188
See also FM response
Index-coupled DFB lasers, 95–97
defined, 25
first-order, 95, 96
highly AR-coated facets, 184
with reflecting facets, 102
second-order, 104–5
special structures, 169–73
threshold modes, 99
See also DFB lasers
Index gratings
current leakage, 136–40

pure, 73
second-order, 79
See also Gratings
Index-guided lasers
carrier injection in, 31, 134–36
lateral current leakage in, 136–40
InGaAsP
active layer, 133
material properties, 126
Injection-locked lasers, 23
InP
material properties, 126
semi-insulating, 128–30
Instantaneous optical frequencies, 81
Intensity modulation (IM). *See* IM response
Interferometers, 179
Fabry-Perot, 180
Mach-Zehnder, 190–91
Michelson, 181, 253
Intermodulation distortion, 219–48
leakage current influence on, 237–40
measurement of, 220
measurement setup, 221
See also Harmonic distortion
Intervalence band absorption (IVBA), 38
Intervalence band transitions, 37

Kirchoff's law, 237
Kramers-Krönig relations, 264

Langevin forces, 69–70, 82, 83, 112
Langevin functions, 33, 35, 74, 84,
112, 255, 272
Large signal
characteristics, rate equation, 52–55
dynamic analysis, 113
properties, 17–18
Laser diodes
characteristics, 15–19
current path possibilities, 131
currents flowing in, 143
device structure, 4–9
differential resistance of, 286
with distributed optical feedback, 23–25
double heterostructure (DH), 3, 5
dynamic characteristics, 17–19
dynamic single-mode (DSM), 21–25
Fabry-Perot (F-P), 2
interest in, 1
large-signal properties, 17–18
lateral index-guiding, 9
lateral optical guiding mechanisms, 8
long-wavelength, parameter values, 152
microwave propagation, 144

Laser diodes (continued)
  noise properties, 17
  operation of, 10–15
  in optical communication systems, 19–21
  optical material parameters, 12–13
  physical processes, 62–64
  P/I curve, 15–16
  rate equation theory, 29–58
  small-signal properties, 19
  spectral properties, 16–17
  static characteristics, 15–17
  thermal aspects of, 13–15
  See also DFB laser diodes
Lateral spatial hole burning, 210–12
  lateral carrier distribution, 211
  rate equation derivation, 210
  See also Spatial hole burning
Leakage current
  emission wavelength and, 239
  as function of bias current, 239
  influence on distortion, 237–40
  on second-order harmonic distortion, 238
LIDAR systems, 251
Linewidth
  calculation, 257
  defined, 251
  of DFB lasers, 259–61
  enhancement factor, 33
    effective, 257, 275
    wavelength vs., 274
  inverse output power vs., 264, 268
  reduction of, 251–52
Linewidth broadening, 261–69
  feedback dispersion, 266–69
  gain suppression, 264–66
  rebroadening due to gain suppression, 265
  side modes, 261–64
Liquid-phase epitaxy (LPE), 279
  disadvantage of, 281
  multibin-boat apparatus, 281
  See also Fabrication
Longitudinal modes, 11
Loss gratings, 88, 99
Low-chirp lasers, 296
Low-gain approximation, 99–100, 105

Mach-Zehnder interferometers, 190–91
Manufacturing costs, 297
Metal-organic chemical vapor deposition (MOCVD). *See* Metal-organic vapor-phase epitaxy (MOVPE)
Metal-organic vapor-phase epitaxy (MOVPE), 283–84
  principle of, 283

  selective-area, 291
  See also Fabrication
Michelson interferometer, 181
  for FM noise detection, 253
  See also Interferometers
Microstrip, 287
Microwave effects, 143–44
Modal gain
  for gratings, 99
  variation of, 86
Mode partition noise, 269, 272
Molecular beam epitaxy (MBE), 279
  gas source (GSMBE), 282
  principle of, 282
  solid-source, 282
  See also Fabrication
Multielectrode lasers
  schematic view of, 168
  tuning, 177
Multiphase-shifted lasers, 4
  fabrication of, 169
  longitudinal power variation, 170
Multi-quantum-well (MQW) DFB lasers, 4
  amplitude modulation response, 214, 215
  material parameters, 225
  See also DFB lasers
Multisection DFB lasers, 4, 17
  division into parts, 108
  with inhomogeneous injection, 296
  See also DFB lasers

Narrowband approach, 106–13
  analysis types, 107
  continuous-wave analysis, 109–11
  large-signal dynamic analysis, 113
  noise analysis, 112–13
  small-signal dynamic analysis, 111–12
  threshold analysis, 108–9
  See also Coupled-mode model;
    Coupled-mode theory
Newton-Raphson (NR) iteration procedure, 109
Noise
  analysis, 112–13
  characteristics, 251–75
  FM. *See* FM noise
  intensity. *See* Relative intensity noise (RIN)
  measuring characteristics of, 252–54
  mode partition, 269, 272
  properties, 17
  in rate equations, 33–36
  shot, 85
Non-return-to-zero (NRZ), 293
Nonuniform injection, 168–69
  multielectrode lasers, 168

power in main and side modes for, 169
Ohm's law, 129
Optical communication systems, 19–21
  direct detection, 19–20
  heterodyne/homodyne, 20–21
  subcarrier multiplexing (SCM), 20
  types of, 19–20
Optical coupling, 287
  alignment tolerances, 290
  efficiency as function of lateral
      misalignment, 290
  efficiency as function of mode width
      ratio, 289
Optical materials, parameters of, 12–13
Optical modulation depth (OMD), 220
Optical power
  average, 110
  buildup, 11, 12
  total density, 170
Optical process, 63
Optical spectrum
  for Fabry-Perot lasers, 150, 151
  threshold for DFB laser, 155
Optical time-domain multiplexing (OTDM), 295
Optical wave propagation, 66–75
  coupled-wave equation reduction, 70–75
  Langevin force, 69–70
  optical field description, 66–68
  scalar wave equation, 68–69
Optoelectronic integrated circuits (OEIC), 25
Organization, this book, 26
Oxide stripe laser, 134

Packaging, 279–80, 285–91
  electrical aspects, 286–87
  optical aspects, 287–91
  price of, 297
  See also Fabrication
Parasitic elements, 140–43
  capacitances, 123
  circuit modeling, 144–45
  depletion capacitance, 140
  differential capacitance, 140
  distributed capacitance, 141
  junction capacitance, 141
  resistances, 123
  series resistance, 140
Phase equations, 32–33
  overview, 45
  See also Rate equations
Photon density rate equations, 31–32
  defined, 31
  overview, 45

total loss of mode, 32
See also Rate equations
Photonic integrated circuit (PIC), 25
Physical processes, 62–64
  electrical process, 62
  electro-optic process, 62–63
  illustrated, 63
  optical process, 63–64
  thermal process, 64
P/I curve, 15–16
  illustrated, 15
  linear, 16
  threshold current, 16
Planar technology, 286
Power spectrum, 55–58
  calculating, 56
  Lorentzian shape of, 57
  from phase fluctuations, 261
Propagator matrix, 95

Quantum-well lasers
  differential transport factor, 214
  dynamics of, 212–16
  SCH, 212–13
Quantum wells, 39–41
  energy levels in, 41
  strained-layer, 41–42
  unstrained, 42
Quasi-continuous tuning, 177

Radiation modes, 76–79
  excited, 77
  Fabry-Perot lasers, 76
Rate equations, 29–58
  carrier density, 30–31
  coupling of, 29
  derivation of, 114–16
  FM/AM behavior, 46–50
  harmonic distortion characteristics, 50–52
  intensity noise, 55–58
  large-signal characteristics, 52–55
  linewidth, 55–58
  longitudinal for optical fields, 79–80
  noise introduction in, 33–36
  overview of, 45
  phase equations and, 32–33
  photon density, 31–32
  power spectrum, 55–58
  solutions of, 44–58
  for spatial hole burning, 196
  static side-mode suppression, 46
Reflecting facets, 101–2, 156
  effect of, 101
  gain suppression and, 228

Reflecting facets (continued)
   single mode operation and, 102
Refractive index
   in bulk materials, 38–39, 40
   carrier dependence of, 163
   gain oscillations and, 18
   for laser waveguides, 68
   perturbation of, 68–69
   temperature and, 14
Relative intensity noise (RIN), 15, 17
   of DFB lasers, 269–73
   frequency dependence, 270–71
   low-frequency, 271–73
   of main mode, 272
   multimode lasers and, 271–73
   single-mode lasers and, 273
   spectrum of single-mode laser, 270
   See also Noise
Relaxation oscillations, 221–24, 271
Resistances, 123
   of cladding layers, 141
   series, 140
   See also Capacitances
Ridge waveguide lasers, 134
Round-trip gain, 108
   for cleaved Fabry-Perot laser, 151
   complex, at threshold, 268
   for DFB lasers, 156
   wavelength dependence of, 154
   See also Gain

Scalar wave equation, 68–69
   decomposing, 72
   scalar field, 71
SCH quantum-well lasers, 212–13
Screening length, 127, 128
Second-order distortion, 51
   in l/4-shifted DFB laser, 230, 231
   from carrier density dependence, 226
   CATV systems and, 248
   current leakage effect on, 238
   in Fabry-Perot laser, 229
   as function of total bis output power, 246, 247
   from intrinsic nonlinearities, 222
   static, AR-coated laser, 233
   static, cleaved laser, 232, 241
   See also Harmonic distortion
Semi-insulating buried heterostructure (SIBH), 138, 140
Semi-insulating InP, 128–30
   I-V characteristics, 130
   Poisson's equation, 129
   properties, 128

Separate confinement heterostructure (SCH), 212–13
Shockley-Read-Hall model, 128
Short-cavity lasers, 21
Shot noise, 85
Side modes
   defined, 12
   far above threshold, 263
   far below threshold, 262
   linewidth and, 261–64
   reaches threshold, 262
Single-mode suppression ratio (SMSR), 147, 159–67
Single quantum well (SQW), 212
Small-signal dynamic analysis, 111–12
   first-order, 112
   higher-order harmonic components, 111
   second-order analysis, 112
   See also Narrowband approach
Small-signal properties, 19
Spatial hole burning, 61, 261
   corrected yield for cleaved lasers, 166, 167
   cutoff frequency, 195–96, 235
   defined, 117
   direct modulation and, 203–4
   elimination of, 171
   in Fabry-Perot lasers, 228
   FM at low bias, 207
   FM contribution at low frequencies, 204
   gain suppression and, 206
   high power level saturation, 164
   impact on side-mode onset, 162, 163
   influence of, 117
   influence on harmonic distortion, 226–36
   internal absorption and, 236
   lateral, 210–12
   longitudinal, 117
   rate equation, 196
   reduced, 167–77
   SMSR degradation by, 159–67
Spectral hole burning, 261
   defined, 42
   interaction via, 261
Spectral properties, 16–17
Spectrum
   device parameter extraction from, 182–84
   FM noise, 253
   measurement, 179–81
   optical, 150, 151, 155
   power, 55–58
Spectrum analyzers, 179
   Fabry-Perot interferometric, 180, 181
   Michelson interferometric-based, 181
Spontaneous emission, 81–83

amplified, 147, 148–55
   one-dimensional model for, 149
   total intensity caused by, 150
Standing-wave effect
   defined, 87
   in gain-coupled lasers, 86–87
Standing-wave factor, 116
Static side-mode suppression, 46
Stepwise constant approximation, 172
   longitudinal variation of power, 174
   longitudinal variation of stepwise constant, 173
Stopband gap, 100–101
Stripe contact DFB laser diodes, 9
Subcarrier multiplexing (SCM), 20
Synchronous digital hierarchy (SDH), 2

TE modes, 67
   guided, 77
   lowest order, 70
Thermal cutoff frequency, 199
Thermal effects, 13–15
   gain and, 14
   refractive index and, 14
   structure-related, 14
Thermal FM, 209
Thermal process, 64
Third-order distortion
   of l/4-shifted laser, 231
   from intrinsic nonlinearities, 223
   static, cleaved laser, 233
   See also Harmonic distortion
Threshold analysis, 108–9
   amplitude/phase resonance conditions, 108
   purpose of, 108
   See also Narrowband approach
Threshold solutions, 94–105
   l/4 phase-shifted DFB laser, 102–4
   reflecting facets, 101–2
   second-order index-coupled DFB lasers, 104–5
   stopband or energy gap, 100–101
   theoretical analysis and, 94
Time-domain-multiplexed (TDM) systems, 294–95
TM modes, 67, 76
Transfer matrix, 153
Tunable twin-guide DFB lasers (TTG), 178, 179
Tuning, 177–79
   continuous, 177
   discrete, 177
   laser alternatives, 179
   quasicontinuous, 177
   typical characteristics of, 178

Vapor-phase epitaxy (VPE), 279
Voltage-driven lasers, 85–86

Wavelength
   calibration, 15
   fine tuning, 17
   tunability, 177–79
   windows, 76
Wavelength-division multiplexing (WDM), 2, 295

Yield, 155–58
   calculated numbers for, 158
   experimental, 158
   gain-coupled laser, 157, 158

The Artech House Optoelectronics Library

Brian Culshaw, Alan Rogers, and Henry Taylor, Series Editors

Acousto-Optic Signal Processing: Fundamentals and Applications, Pankaj Das

Amorphous and Microcrystalline Semiconductor Devices, Volume II: Materials and Device Physics, Jerzy Kanicki, editor

Bistabilities and Nonlinearities in Laser Diodes, Hitoshi Kawaguchi

Chemical and Biochemical Sensing With Optical Fibers and Waveguides, Gilbert Boisdé and Alan Harmer

Coherent and Nonlinear Lightwave Communications, Milorad Cvijetic

Coherent Lightwave Communication Systems, Shiro Ryu

Elliptical Fiber Waveguides, R. B. Dyott

Field Theory of Acousto-Optic Signal Processing Devices, Craig Scott

Frequency Stabilization of Semiconductor Laser Diodes, Tetsuhiko Ikegami, Shoichi Sudo, Yoshihisa Sakai

Fundamentals of Multiaccess Optical Fiber Networks, Denis J. G. Mestdagh

Germanate Glasses: Structure, Spectroscopy, and Properties, Alfred Margaryan and Michael A. Piliavin

Handbook of Distributed Feedback Laser Diodes, Geert Morthier and Patrick Vankwikelberge

High-Power Optically Activated Solid-State Switches, Arye Rosen and Fred Zutavern, editors

Highly Coherent Semiconductor Lasers, Motoichi Ohtsu

Iddq Testing for CMOS VLSI, Rochit Rajsuman

Integrated Optics: Design and Modeling, Reinhard März

Introduction to Lightwave Communication Systems, Rajappa Papannareddy

Introduction to Glass Integrated Optics, S. Iraj Najafi

Introduction to Radiometry and Photometry, William Ross McCluney

Introduction to Semiconductor Integrated Optics, Hans P. Zappe

Laser Communications in Space, Stephen G. Lambert and William L. Casey

Optical Fiber Amplifiers: Design and System Applications, Anders Bjarklev

Optical Fiber Communication Systems, Leonid Kazovsky, Sergio Benedetto, Alan Willner.

Optical Fiber Sensors, Volume Two: Systems and Applicatons, John Dakin and Brian Culshaw, editors

Optical Fiber Sensors, Volume Three: Components and Subsystems, John Dakin and Brian Culshaw, editors

Optical Fiber Sensors, Volume Four: Applications, Analysis, and Future Trends, John Dakin and Brian Culshaw, editors

Optical Interconnection: Foundations and Applications, Christopher Tocci and H. John Caulfield

Optical Network Theory, Yitzhak Weissman

Optoelectronic Techniques for Microwave and Millimeter-Wave Engineering, William M. Robertson

Reliability and Degradation of LEDs and Semiconductor Lasers, Mitsuo Fukuda

Reliability and Degradation of III-V Optical Devices, Osamu Ueda

Semiconductor Raman Laser, Ken Suto and Jun-ichi Nishizawa

Semiconductors for Solar Cells, Hans Joachim Möller

Smart Structures and Materials, Brian Culshaw

Ultrafast Diode Lasers: Fundamentals and Applications, Peter Vasil'ev

For further information on these and other Artech House titles, contact:

Artech House
685 Canton Street
Norwood, MA 02062
617-769-9750
Fax: 617-769-6334
Telex: 951-659
email: artech@artech-house.com

Artech House
Portland House, Stag Place
London SW1E 5XA England
+44 (0) 171-973-8077
Fax: +44 (0) 171-630-0166
Telex: 951-659
email: artech-uk@artech-house.com

WWW: http://www.artech-house.com